Springer Proceedings in Earth and Environmental Sciences

The series Springer Proceedings in Earth and Environmental Sciences publishes proceedings from scholarly meetings and workshops on all topics related to Environmental and Earth Sciences and related sciences. This series constitutes a comprehensive up-to-date source of reference on a field or subfield of relevance in Earth and Environmental Sciences. In addition to an overall evaluation of the interest, scientific quality, and timeliness of each proposal at the hands of the publisher, individual contributions are all refereed to the high quality standards of leading journals in the field. Thus, this series provides the research community with well-edited, authoritative reports on developments in the most exciting areas of environmental sciences, earth sciences and related fields.

James T. Anderson
Editor

Advances in Environmental Sustainability

Proceedings of 2022 8th International Conference on Advances in Environment Research

Springer

Editor
James T. Anderson
Clemson University
Georgetown, SC, USA

ISSN 2524-342X ISSN 2524-3438 (electronic)
Springer Proceedings in Earth and Environmental Sciences
ISBN 978-3-031-26364-4 ISBN 978-3-031-26365-1 (eBook)
https://doi.org/10.1007/978-3-031-26365-1

This Springer imprint is published by the registered company Springer Nature Switzerland AG
The registered company address is: Gewerbestrasse 11, 6330 Cham, Switzerland

Preface

The 2022 8th International Conference on Advances in Environment Research (ICAER 2022) was held successfully during April 22–24, 2022. The conference fosters communication among researchers and practitioners working in various scientific areas with a common interest in improving advances in environmental research. Many researchers, engineers, academicians, and industry professionals worldwide presented their research results and development activities.

This conference was initially to be hosted offline in Singapore. However, due to the spread of COVID-19, the complexities of the pandemic among the many countries involved, and the strict entry-exit management of the local government, the conference committee decided to hold ICAER 2022 as a virtual conference. In the context of the normalization of the epidemic, ICAER 2022 online mode can effectively guarantee safety, arouse the enthusiasm of participants, and increase attendance due to the restrictions of the pandemic.

The ICAER 2022 proceedings are a collection of outstanding submissions from universities, research institutes, and industries. The papers were peer-reviewed by conference committee members and international reviewers. The manuscripts selected depended on their quality and their relevancy to the conference. This volume intends to present advances in environmental research and related areas, such as environmental science and technology, environmental dynamics, global environmental change and ecosystems, soil decontamination, environmental sustainability, health and the environment, and environmental dynamics.

We express our deepest gratitude to all authors for their effort in preparing papers. We thank the organizing committee, reviewers, speakers, chairpersons, and sponsors for their valuable advice in the organization and helpful peer review of the papers.

Georgetown, SC, USA James T. Anderson

Contents

Conference Committees

Conference Chairs
Kokyo Oh, Center for Environmental Science in Saitama, Japan
Ngai Weng Chan, Universiti Sains Malaysia, Malaysia

Program Chairs
Kei EGUCHI, Fukuoka Institute of Technology, Japan
Plamena Zlateva, Bulgarian Academy of Sciences, Bulgaria

Steering Committee
Eric James Strauss, Michigan State University, East Lansing, USA

Technical Committee
J. Helan Chandra, Jeppiaar Engineering College, India
Bashir Ahmed Mir, National Institute of Technology Srinahgar, Kashmir, J&K, India
Radu Godina, FCT, NOVA University Lisbon, Portugal
Dimiter Velev, University of National and World Economy, UNSS, Bulgaria
Plamena Zlateva, Institute of Robotics - Bulgarian Academy of Sciences, Bulgaria
Nour Mohammad, Premier University, Chittagong, Bangladesh
Mamta Sharma, Raj Rishi Bhartrihari Matsya University, India
Beibei Li, Nanjing University, China
Myronidis Dimitrios, Aristotle University of Thessaloniki, Thessaloniki, Greece
M.E. Raghunandan, Monash University, Malaysia
Ahmed Jalal Khan Chowdhury, International Islamic University Malaysia, Malaysia
Kei EGUCHI, Fukuoka Institute of Technology, Japan
Haliza Abdul Rahman, Universiti Putra Malaysia (UPM), Malaysia
Shirley C. Agrupis, Mariano Marcos State University, City of Batac, Philippines
Eric James Strauss, Michigan State University, East Lansing, USA
Ahmed Jalal Khan Chowdhury, International Islamic University Malaysia, Malaysia
Wenchuan Ding, Chongqing University, China
Chan Jin Park, Incheon National University, South Korea

T. S. Devaraja, Philosophy in Commerce, University of Mysore, India
Md. Nazrul Islam, Jahangirnagar University, Bangladesh
M. Azizul Moqsud, Yamaguchi University, Thailand

Part I

Environmental Pollution Control and Renewable Energy Technology

Tree Bark as a Bioindicator of Atmospheric Contamination by Heavy Metals According to Vehicular Traffic Intensity in El Tambo, Huancayo, Peru

Neddy Milka Baltazar Sedano,
Andrea Jesús Schwartz Valverde,
Alexis Jeanpeer Guerreros Chiri,
and Steve Dann Camargo Hinostroza

Abstract

Tree bark is an excellent bioindicator of atmospheric heavy metal contamination. This study aims to determine heavy metals from tree bark in El Tambo, Huancayo, Peru. *Salix babylonica*, *Populus nigra*, *Senna multiglandulosa*, and *Schinus molle* bark samples were taken in 2021 and analyzed through Inductively Coupled Plasma Optical Emission Spectroscopy. Results show heavy metals concentrations ($mg.kg^{-1}$) of Zn (195.92 ± 125.97) > Pb (24.45 ± 15.57) > Cu (23.39 ± 11.01) > Cr (2.43 ± 1.13) > Cd (1.77 ± 3.06) > Ni (1.01 ± 0.70). The highest contamination levels were by Zn, Pb, and Cu, and the lowest by Cr, Cd, and Ni. The average concentration due to vehicular traffic intensity was Zn > Cu > Pb > Cr > Cd > Ni. Cr reveals higher concentration at high traffic levels; Cu, Ni, Pb, and Zn at moderate, and Cd at low levels. *Schinus molle* showed the highest concentration of Ni, *Populus nigra* of Cr, and *Salix babylonica* of Zn and Cd. The lowest concentrations of all metals were found in *Senna multiglandulosa*. This and *Salix babylonica* are bioindicators of Pb, Cr, Cd, Cu, Ni, and Zn, the first report for the world on bark in Fabaceae and Salicaceae species.

Keywords

Air pollution · Bark · Heavy metals · Indicator organisms · Traffic

N. M. Baltazar Sedano (✉) · A. J. Schwartz Valverde · A. J. Guerreros Chiri
Faculty of Environmental Engineering, Universidad Continental, Huancayo, Peru

S. D. Camargo Hinostroza
Research Unit of the Faculty of Environmental Engineering, Universidad Continental, Huancayo, Peru

J. T. Anderson (ed.), *Advances in Environmental Sustainability*, Springer Proceedings in Earth and Environmental Sciences,
https://doi.org/10.1007/978-3-031-26365-1_1

1 Introduction

Air pollution in large cities comes from various sources, with vehicles being the primary source of emissions and associated with vehicular traffic intensity. The latter, in turn, concerns the flow of light or heavy vehicles, the type of fuel, the age of the vehicles, their maintenance, the state of the vehicles, and their speed [1]. Heavy vehicular traffic produces heavy metals that accumulate in the atmosphere [2]. Pollutants emitted by vehicles include heavy metals potentially toxic to health, such as lead, cadmium, and zinc, among others. In addition, heavy metal pollution is potentially a persistent problem worldwide in increasing order [3, 4]. Weather conditions and local sources play an essential role in the concentrations of particulate matter in the air that are important carriers of metals [5].

Detecting trace metals deposition, accumulation, and distribution in ecosystems can be done through biomonitoring. This method employs various tree organs or parts [6] to detect in the atmosphere trace metal elements [7]. Besides, lichens have proven to be excellent biomonitoring organisms due to their homogeneous geographic distribution, slow growth, great capacity to accumulate elements, and easy sampling compared to other biomonitors [8]. The use of plant bioindicators has many advantages. The most significant are low cost, long-term sampling, and high availability [9]. The use of phytoindicators is increasingly being used to assess the quality of an ecosystem due to their sensitivity to chemical variations in the composition of the environment and the accumulation of pollutants. Additionally, atmospheric trace metal concentrations have been successfully monitored utilizing several vegetation types [6, 10, 11].

Tree bark is a tool to measure the consequence of heavy metal pollution and monitor air quality [12]. Moreover, it allows for identifying and mapping organic and inorganic pollutants and accumulating atmospheric particles due to its absorbent surface and deeply grooved structure [13]. The use of tree bark as a biomonitoring technique was established to estimate the levels of atmospheric contamination by heavy metals in urban areas [14, 15]. Additionally, tree bark retains higher trace metal concentrations than air and rainwater [6, 16].

The inorganic content of heavy metals in the atmosphere measured in tree bark is a surface deposit that should not be considered a simple cumulative material but rather an excellent source of information that points to recent geotechnical and anthropogenic pressures [14].

In large cities worldwide, air pollutant emissions are generated by vehicular traffic [17]. The sources of toxic elements such as Zn, Cd, Cu, Pb, Ni, and Cr are emissions released by vehicles on the surface and street dust from the ground [18, 19]. Traffic-related metal emissions come from vehicle exhaust pipe emissions from burning fuels, lubricating oils, and engine wear. These, in turn, depend on oil and fuel compositions. Gasoline-fueled vehicles have been associated with the highest emissions of Cu and Ni.

The city of Huancayo in Peru shows a significant air quality problem due to particulate matter (PM) as the primary source of contamination. In 2017, PM_{10} and $PM_{2.5}$ were much higher than the annual Peruvian air quality regulation standard. Furthermore, even $PM_{2.5}$ is a much higher value than those registered in other Peruvian cities. The vehicle fleet is the most critical emission source as it causes a greater risk to the health of the inhabitants [20]. The estimated concentrations of air pollutants by vehicles with gasoline- and diesel-fueled engines exceed Peru's Environmental Quality Standards (EQS), especially carbon monoxide [21].

Sauce llorón or weeping willow (*Salix babylonica*) belongs to the Salicaceae family. It has high biomass productivity, high transpiration rates, and great potential for accumulation and tolerance to Pb, providing valuable and scientific information for phytoremediation research in places with trees under Pb stress [22].

Álamo or black poplar (*Populus nigra*) is a deciduous perennial tree species of the Salicaceae family. It is characterized by its rapid growth, extensive root system, high biomass production, high resistance to water and nutrient scarcity, to storms and cold, and it has a high tolerance and potential for the accumulation of metals [23, 24]. It accumulates metals through its root system and foliage and translocates and stores high Cd contents in the leaves and bark [25, 26].

Molle or Peruvian or American pepper (*Schinus molle*) belongs to the Anacardiaceae family. This species has been utilized in San Luis de Potosí, Mexico, to evaluate the environmental impact of different land uses (agricultural, rural, urban, residential, commercial, and mining). For this, the concentrations of Al, As, Co, Cu, Cd, Pb, Ti, V, and Zn were analyzed in leaflets [27].

In the Mantaro Valley in the Andean region of Peru, the "Millhua tanquish" shrub or glandular senna (*Senna multiglandulosa*) is an adapted species [28] that belongs to the Fabaceae family, and is appreciated and valued as fuel and as an ornamental in the main green areas of the city of Huancayo [29–31].

Plants with an exceptional ability to accumulate metals in the plant kingdom are called hyperaccumulators [32]. Its discovery worldwide has been slow due to the lack of systematic screening of plant species in various regions [33]. In 2017, a global database of plants that hyperaccumulate metals and metalloids included the species *Senna obtusifolia* [34].

According to the Agency for Toxic Substances and Disease Registry (ATSDR), Pb is found in small amounts in the earth's crust and throughout the environment. Lead is produced through human activities such as burning fossil fuels. Cadmium is naturally found on the earth's crust. It adheres firmly to the organic matter in which it remains immobile in the soil and can be incorporated by plants. Chromium is present in the environment in various forms. Copper is found naturally throughout the environment. It can enter the environment from landfills, domestic wastewater, the combustion of waste and fossil fuels, wood and phosphate fertilizers production, and natural sources such as dust in the air, ground, volcanoes, decaying vegetation, and forest fires. Zinc is one of the most common elements in nature [35].

Inductively Coupled Plasma Optical Emission Spectroscopy (ICP-OES) is an appropriate technique to determine metal elements [36]. These authors state that this methodology is very promising for environmental studies since various elements can be measured and are highly sensitive to detecting heavy metals.

Studies that use biomonitors to measure environmental pollutants in Peru are very scarce. The district of El Tambo at 3260 m a.s.l. is densely populated, where the primary source of heavy metal contaminants is vehicular traffic. The primary contaminant elements found are Al, Ca, Fe, K, and Na, as well as trace elements such as As, Ba, Cd, Ce, Co, Cr, Cu, La, Mn, Ni, Pb, Sb, Sc, Sr, U, V, and Zn [37].

Peru has little scientific knowledge about the bioaccumulation of heavy metal in the bark of *Salix babylonica, Populus nigra, Senna multiglandulosa,* and *Schinus molle* as a consequence of atmospheric contamination. Hence, it is necessary to contribute to the knowledge of the environmental state of El Tambo District, Huancayo Province, Junín Region of Peru, considering the ecological importance and the absence of quantitative records.

Accordingly, this study aims to evaluate the bark of the species *Salix babylonica, Populus nigra, Senna multiglandulosa,* and *Schinus molle* as bioindicators of atmospheric contamination by traces of heavy metals determined by the intensity of vehicular traffic in the main urban areas with ornamental trees in the District of El Tambo, Huancayo, Peru.

2 Materials and Methods

2.1 Study Area, Vehicular Traffic Intensity, and Species

This study was carried out in the district of El Tambo, province of Huancayo, Junín Region, Peru (Fig. 1). It has a population of 166,359 inhabitants [38] and is located at 3260 m a.s.l. It is densely populated, where the main source of pollution is vehicular traffic comprised of vehicles such as massive transport, rural pickups (5% and 20%), motorcycles (3%), public transport buses (4–5%), collective cars, independent taxis, company taxis, school service vehicles, loading, and unloading vehicles and interprovincial buses (4–6%), for a total of 11,135 vehicles [39, 40]. There are about 49 vehicles per thousand inhabitants [41].

The climate of the study area is temperate, with a maximum temperature of 15 °C in December and a minimum of 1 °C in July. It is characterized by its short and cloudy summers, and the rest of the year is dry (MINAM, 2022). It had a maximum rainfall of 121 mm.

The district of El Tambo has high vehicular traffic congestion that causes air pollution due to city expansion. In addition, suburban development and low maintenance of roads and paths are observed benefitting middle to upper-class car users [42]. The intensity of the vehicular traffic considered in this study (Table 1) was the one occurring solely on main roads where automotive vehicles circulate [20, 37, 43].

Bark samples were taken from trees near streets with different levels of traffic intensity. The trees sampled belong to the species *Populus nigra* (Pni), *Salix*

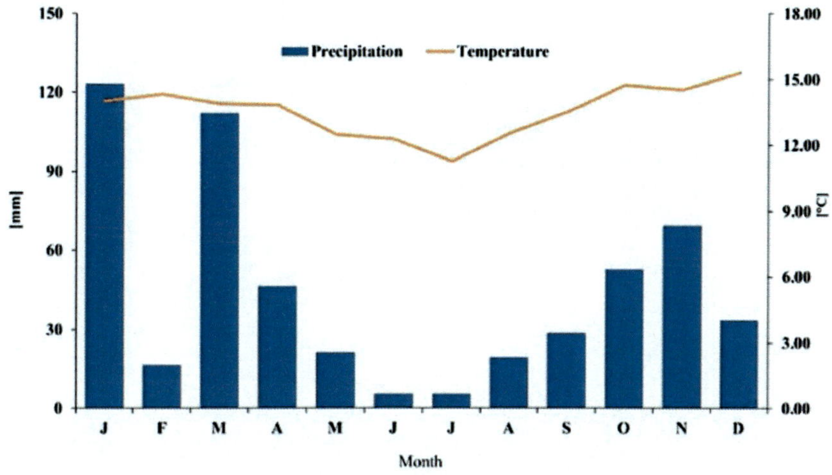

Fig. 1 Climatogram of the District of El Tambo, Huancayo, year 2021

Table 1 The intensity of vehicle traffic

Location	Scale
(a) Intihuatana, Ciudad Universitaria, Caminito de Huancayo, Parque Infantil	High
(b) Parque Industrial y Parque El Triangulo	Moderate
(c) Parque de los Sombreros y Umuto	Low

babylonica (Sba), *Schinus molle* (Smo), and *Senna multiglandulosa* (Smu) (Fig. 2), as these are the most predominant species established as ornamental plants in the main streets and avenues of the District of El Tambo [1].

2.2 Sampling and Bark Preparation

During the first quarter of 2021, we collected bark samples from 80 trees (20 trees/species) of *Populus nigra*, *Salix babylonica*, *Schinus molle*, and *Senna multiglandulosa*. Five bark samples were collected for each species at a height of 1.5 m from the soil surface. The amount of bark collected was 100 g and stored in a cooler container at 4 °C ± 0.1 for 24 h. The impurities of the outer bark were cleaned, removing very mature cells with the help of a brush, a paintbrush, and a fine thread brush. Next, the bark was ground manually in a porcelain mortar, homogenizing the bark particles and obtaining 50 g of sample. These samples were dried in an Inucell oven at 60 °C ± 0.1 for 90 h until constant weight, and the final anhydrous moisture content was determined. The bark was then ground with a Pulviressette mill to a particle diameter of 50 microns. Finally, we weighed 5 g of bark from each sampled tree and sent it to the laboratory for further analysis.

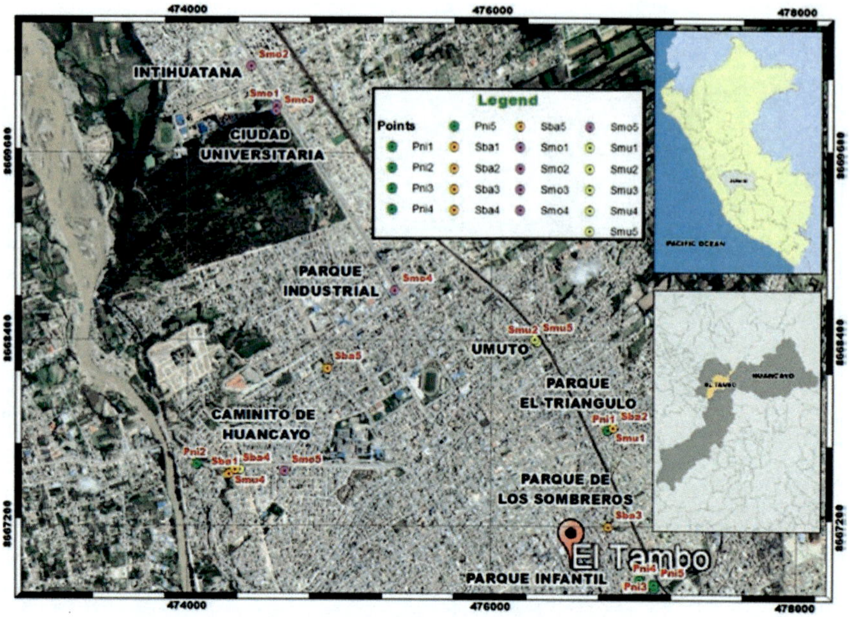

Fig. 2 Study area map of the District of El Tambo, Huancayo, Peru, with the location of the sampling points. Pni (green dots): *Populus nigra,* Sba (red dots): *Salix babylonica*, Smo (pink dots): *Schinus molle*, Smu (orange dots): *Senna multiglandulosa*. (Source: Elaborated by the authors)

2.3 Laboratory Analyses

Please Digestion was carried out using a sample of 0.5 g of dry bark, a mixture of nitric acid 65% p.a. (EMSUR® Reag. Ph Eur, ISO), and hydrochloric acid (EMSURE® ACS, ISO, Reag. Ph. Eur. ISO) were added. Digestion initiated at a temperature of 85 °C for 4 h. It was then cooled and filtered through a Whatman® quantitative ashless, Grade 40 filter paper into a 50 mL volumetric flask. The digested solution was read in an Inductively Coupled Plasma Optical Emission Spectrometer (ICP-OES) (Agilent, Model 700). Results were processed with the ICP-EXPERT program series version 4.1.0 b 443, registering heavy metal concentration values in mg kg^{-1}. The readings of each metal concentration were fitted to the calibration curves with a value of r = 1 (Table 2). The standardized EPA 200.7 method, "Determination of Metals and Trace Elements in Water and Wastes by Inductively Coupled Plasma-Atomic Emission Spectrometry," was used (U.S. EPA, 1994).

The instrumental conditions of the ICP-OES method were Power (kW) = 1.30; Plasma flow (L/min) = 16.5; Auxiliary flow (L/min) = 1.50; Nebulizer flow (L/min) = 0.75; Replica reading time (s) = 10, and Instrument Stabilization Delay (s) = 15. The sample introduction parameters were Sampling Delay (s) = 55, Pump Speed (rpm) = 5, and Wash Time (s) = 5.

Table 2 Major and trace element calibration standards

Element (mg kg^{-1})	Pattern	Concentration (mg L^{-1})	Intensity	Equation	r2
Cd	1	0	13.0217	Y = 10435x – 24.392	0.96
	2	0.2	2020.95		
	3	2	20848.9		
Cr	1	0	3.13641	Y = 1019x + 6.2257	0.96
	2	0.5	519.17		
	3	5	5101		
Cu	1	0	109.504	Y = 17784x + 48.886	0.99
	2	0.2	3538.31		
	3	2	35623.4		
Ni	1	0	7.23193	Y = 506.36x + 4.0786	0.99
	2	0.2	101.847		
	3	2	1017.15		
Pb	1	0	6.04907	Y = 217.7x – 3.4727	0.99
	2	1	203.649		
	3	10	2174.6		
Zn	1	0	4.49847	Y = 781.74x + 0.7429	0.97
	2	0.5	387.441		
	3	5	3909.87		

2.4 Statistical Analyses

The data collected were analyzed using multivariate, inferential, and descriptive statistics. One-way analysis of variance (ANOVA) ($p \leq 0.05$) was performed to test differences in heavy metal accumulation by vehicular traffic intensity and per tree species. The analyses were performed with the SPSS v25 software.

3 Results and Discussion

The general average concentrations of heavy metals in the bark of four tree species are presented in Table 3, showing the following descending order Zn > Pb > Cu > Cr > Cd > Ni. The results indicate the presence of Zn, Pb, and Cu in high concentrations, while Cr, Cd, and Ni are registered in low concentrations.

The metals quantified are emitted into the atmosphere mainly by vehicles [44]. Pb and Zn contents correlate with traffic volume [45], and the leading cause of Pb, Cd, Cu, and Zn contamination are automobile tires and brakes. These can adversely affect the atmospheric environment and ecosystems [46]. Likewise, the sites with different anthropogenic activities reveal higher concentrations for Pb, Zn, and Cu, while the concentrations of Cd are lower. This could be due to adding these metals in minimal quantities into petrol to increase gasoline octane [2].

According to the different vehicular traffic intensities, the concentration of heavy metals registered in tree bark (ANOVA, Table 4) did not show significant statistical

Table 3 The average concentration of heavy metals in the bark of four ornamental tree species

Element (mg kg^{-1})	Mean standard deviation (mg kg^{-1})
Cd	1.77 ± 3.06
Cr	2.43 ± 1.13*
Cu	23.39 ± 11.01*
Ni	1.01 ± 0.70
Pb	24.45 ± 15.57*
Zn	195.92 ± 125.97*

(*) Shapiro-Wilk test ($p \geq 0.05$)
*Normality test n = 20, Shapiro-Wilk ($p > 0.05$), Lilliefors Significance Correction

Table 4 ANOVA of heavy metal concentration in tree bark according to traffic intensity in the district of El Tambo

Element (mg kg^{-1})	High	Moderate	Low	P-value 0.05
Cd	0.96 ± 0.88*	2.28 ± 3.28	3.02 ± 5.84	0.514*
Cr	2.57 ± 0.93*	2.55 ± 0.81	1.91 ± 1.99*	0.181
Cu	23.64 ± 9.60*	26.20 ± 9.65*	18.55 ± 16.98*	0.256
Ni	1.02 ± 0.63*	1.12 ± 0.81*	0.85 ± 0.86*	0.244
Pb	23.83 ± 11.25*	26.76 ± 14.00*	22.54 ± 28.51*	0.076
Zn	194.02 ± 93.35*	233.87 ± 119.77*	143.73 ± 209.58	0.569

* Shapiro-Wilk ($p > 0.05$), Lilliefors Significance Correction. P-value ($p > 0.05$)

differences for Cd, Cr, Cu, Ni, Pb, and Zn. However, the mean concentration value in descending order is Zn > Cu > Pb > Cr > Cd > Ni. In addition, Cr reveals a higher concentration at a high traffic intensity. In turn, Cu, Ni, Pb, and Zn show a higher concentration in moderate traffic intensity, and Cd shows a higher accumulation when there is low traffic intensity.

There are three common sources of metallic elements directly related to the intensity of vehicular traffic, including the wear of the brake disk coating that liberates Pb, Zn, Cr, Ni, and Cd; tire wear generates Pb, Zn, Cu, and Cd; and oil leaks produce Pb, Zn, Cu, Ni and Cd [47]. The study revealed that vehicular traffic in the district of El Tambo at peak hours is characterized by having a very high flow on the green areas of main avenues where *S. molle, P. nigra, S. babylonica,* and *S. multiglandulosa* are found.

A study on *Tillandsia capillaris* in the District of El Tambo showed that the average concentration of atmospheric heavy metals pollutants according to the intensity of vehicular traffic showed very low values for Cd, Cr, Cu, and Zn. In contrast, the ones for Ni were very high [37]. However, the current study found the contrary, where the concentrations in the bark in mg kg^{-1} (Table 3) show very high values for Cd, Cr, Cu, and Zn and very low values for Ni. This may be because tree bark is a better indicator and bioaccumulator when monitoring atmospheric pollutants by heavy metals in urban areas [6, 12, 13, 15, 16].

Table 5 ANOVA of heavy metal concentration in the bark of four ornamental tree species

Element (mg kg^{-1})	Tree species				
	S. molle	*P. nigra*	*S. babylonica*	*S. multiglandulosa*	(p-value = 0.05)
Cd	0.34 ± 0.11*	0.95 ± 0.32*	5.63 ± 4.36*	0.15 ± 0.12*	0.003
Cr	2.51 ± 1.10*	2.80 ± 0.49*	2.49 ± 1.42*	1.93 ± 1.46*	0.701**
Cu	27.90 ± 8.92*	26.28 ± 10.17*	22.59 ± 13.01*	16.79 ± 11.50*	0.416**
Ni	1.52 ± 0.65*	0.89 ± 0.51*	0.99 ± 0.98	0.66 ± 0.41*	0.262
Pb	30.88 ± 14.58*	27.09 ± 9.38*	24.98 ± 23.44*	14.85 ± 11.23	0.435**
Zn	132.09 ± 55.39*	226.36 ± 55.29*	361.24 ± 72.97*	63.99 ± 41.86*	0.000**

(*) Shapiro-Wilk ($p \geq 0.05$) (**) and Levene ($p \geq 0.05$) tests

According to the ANOVA (Table 5), there are no significant statistical differences ($p > 0.05$) between the four species concerning the concentration of Pb, Cr, Cu, and Zn. However, there are significant statistical differences ($p < 0.05$) in the concentration of Cd and Ni. Also, each species' average values of heavy metals revealed differences in accumulation. It also shows that the species *Schinus molle* (Molle) has a higher concentration of Ni; *Populus nigra* (black poplar) of Cr; *Salix babylonica* of Zn and Cd; and *Senna multiglandulosa* exhibited lower concentrations of metals such as Cd, Cr, Cu, Ni, Pb, and Zn. The barks of the studied species are confirmed to be good bioaccumulators of metals that come from atmospheric pollution, which, in turn, are related to the intensity of vehicular traffic in the El Tambo district.

The concentration of heavy metals in *S. molle* bark (Table 4) decreased in the following order Zn > Pb > Cu > Cr > Ni > Cd. In addition, [48] indicates that *S. molle* is a good accumulator bioindicator of Cd, Cr, Ni, and Pb, whose Cr concentration is lower and higher in Cd and Pb. At the same time, Ni shows an extremely high concentration compared to what was found in the study. However, the work of [27] showed higher concentrations of Cd, Cu, Pb, and Zn in the leaflet of *S. molle* in San Luis de Potosí, Mexico. The emission sources come from agricultural soil use systems and rural, commercial, and service settlements, indicating the presence of these heavy metals in the bark of *S. molle*. Therefore, it is a good indicator of environmental contamination. The research reveals that the bark of *S. molle* is a good bioindicator of Cd, Cr, Cu, Ni, Pb, and Zn from atmospheric pollution on roads with high, moderate, and low vehicular traffic, and this may be due to the influence of the impact of the land uses to which this species is exposed, such as the use of ornamental trees in main avenues and streets.

The bark of *álamo* or black poplar (*Populus nigra*) revealed heavy metal concentrations in the following order: Zn > Pb > Cu > Cr > Cd > Ni. This species is considered a good bioindicator of atmospheric contamination by heavy metals (mg kg^{-1}) such as Cu (11.360 ± 1.5), Pb (8.627 ± 1.5), Zn (32.308 ± 2.8), and Cd (0.0362 ± 0.02) in Navodari, Romania [49]. Black poplar bark is considered a passive monitor of air pollution by Pb (9.22 ± 0.26 mg kg^{-1}) in Tirupati, India [50]. Also, [51] in Greece found (ppm) Cd (0.137), Cu (0.948), Cr (0.367), and Zn (1.893) in the leaves of this species. In addition, in Turkey, [52] reported concentrations (ug g^{-1}) of Ni (0.484), Pb (0.57), and Zn (0.504). Moreover, [53] registered (ug g^{-1}) Cu (2.27–9.66), Ni (0.84–2.59), and Pb (0.40–0.75) in *P. nigra* leaves. Since there is little information on the transfer of metals from the bark to the leaves, it can be affirmed that the bark of this species is an excellent bioindicator not only for air pollution but also for soil contamination.

The weeping willow (*S. babylonica*) revealed metal concentrations in the following order: Zn > Pb > Cu > Cd > Cr > Ni, being the first record of heavy metals in the bark of this species in El Tambo District, Huancayo Province, Peru. It should be mentioned that, in Ankara, Turkey, [54] studied leaves as bioindicators and bioaccumulators (mg kg^{-1}) of Cd (0.084), Cr (0.024), Cu (0.049), Ni (0.012), Pb (0.008), and Zn (0.014) in different vehicular traffic areas. This species has high biomass productivity and transpiration rates, and its roots are considered excellent Pb phytoextractors [22]. Moreover, it has high resistance to Pb toxicity. Moreover, this species is preferred for its phytoremediation quality of trace metal contaminated sites due to its easy propagation and cultivation, rapid growth, and deep root system, making it worthwhile to mitigate air pollution in green urban areas. It is an efficient phytoextracting species to accumulate Cd [55]. It can remove heavy metals by natural phytoremediation and induced phytoremediation [56]. However, no study has been carried out on the determination of metals in the bark of these ornamental tree species established in urban areas whose streets have high traffic intensity.

S. multiglandulosa showed significant low concentrations (mg.kg^{1}) for Zn (63.99 ± 41.86) and Cd (0.15 ± 0.12) and was not significant for Cu (16.79 ± 11.50), Pb (14.85 ± 11.23), Cr (1.93 ± 1.46) and Ni (0.66 ± 0.41), compared to S. *molle*, *P. nigra* and *S. babylonica* and according to the ANOVA (Table 4). This study reveals that the bark of *Millhua tanquish* (*S. multiglandulosa*) is a good indicator and bioaccumulator of traces of heavy metals in ornamental trees established in green areas with different vehicular traffic intensities in Huancayo.

Different concentrations of Zn, Pb, Cu, Cd, Cr, and Ni in the bark of the four tree species studied suggest they are good accumulators of heavy metals in different urban vehicular traffic intensity areas. In particular, this is due to the various places from which the bark samples were extracted since, in many Latin American cities, the traffic conditions of vehicles are intense due to inappropriate handling by the authorities. Thus, in the Junín region, Huancayo, El Tambo, the conditions are not different from this reality. Considering that the species studied have been established for more than 15 years in the main areas of the district of El Tambo and that in the Junín region, the sale of gasoline with lead was prohibited as of September 1, 2007 [57], it is stated that one of the primary sources for the presence of Pb is related to fuel consumption and the high level of traffic in the district of El Tambo. Moreover, it is probably also due to the bioaccumulation generated by emissions from the DOE RUN metallurgical plant complex of La Oroya, which come from lead circuit operations carried through the air due to favorable weather conditions accumulating in the soil. The wind suspended these particles and reached human activities [58]. This result reveals that the atmospheric and environmental contamination by traces of Zn, Pb, Cu, Cd, Cr, and Ni is silent and seriously affects the health of the inhabitants [59]. Therefore, controlling the contamination sources and performing toxicity tests are necessary [60]. Moreover, the results in the five species exceed the lead concentration levels in El Tambo, which are higher than Pb <0.5 ug m^{-3} for air quality [61].

Pesticides, fertilizers, and other chemical components used in the agricultural farmwork of the district of El Tambo contaminate the soil due to the heavy metals and other toxic elements they contain. Rainwater and surface runoff infiltrates green areas, transporting harmful ions to the soil in solution until they are deposited in aquifers [62] and absorbed by tree roots. Considering that these pollutants do not reach the plants only through the air but also through the soil, the Cd and Pb results do not exceed the Environmental Quality Standards for residential land or parks [63]. Nevertheless, in relation to Cr, the value is higher. This is probably because this element is found naturally in rocks, animals, plants, and the soil of the district of El Tambo.

The bark of the species *Salix babylonica, Populus nigra, Senna multiglandulosa,* and *Schinus molle* exposed to air pollution in the El Tambo district are natural absorbers [64], providing crucial, precise, and valuable information on the presence of heavy metals such as Zn, Pb, Cu, Cd, Cr, and Ni, evidencing the changes in the conditions through the biomonitoring carried out. In addition, *Senna multiglandulosa* and *Salix babylonica* are bioindicators of heavy metals such as Zn, Pb, Cu, Cd, Cr, and Ni; this study provides the first report on bark and heavy metals for the Fabaceae and Salicaceae families for Latin America and the world. Likewise, the species *Schinus molle* can be considered a potential bioindicator of heavy metals in urban areas and the Peruvian Andes, where this species has a wide distribution. It is viewed as a component of agroforestry systems. This will allow reducing the levels of soil and air contamination by heavy metals in Andean agricultural systems. With the results found in this study, better handling of the current critical environmental situation through sustainable forest management can improve the highly polluted urban environment of different anthropogenic sources. Furthermore, these can also be used to establish an environmental management baseline to enhance the life quality of the inhabitants of the district of El Tambo.

4 Conclusions

The bark of *Salix babylonica, Populus nigra, Senna multiglandulosa,* and *Schinus molle* are bioindicators of heavy metals in green areas according to vehicular traffic intensity in the district of El Tambo. The evaluated heavy metals Cd, Cr, Cu, Ni, and Pb are mainly emitted into the atmosphere by vehicle traffic. The high concentrations are associated with fossil fuels and tire wear. For Pb, it is probably due to the convection flow of ambient passive air generated by the metallurgical activity of DOE RUN – La Oroya. It is evident that there is little knowledge, and there is a need to deepen our understanding of *Senna multiglandulosa* and *Salix babylonica* as potential bioaccumulators of heavy metals.

Acknowledgments To Hernán Baltazar Castañeda, Ph.D., Research Professor of Universidad Nacional del Centro del Peru (UNCP) for his valuable contribution in the statistical analysis and writing of the manuscript. Many thanks to the Faculty of Forestry and Environmental Sciences of UNCP for granting us access to the Biodiversity and Forest Management Laboratory equipment.

References

1. Moreira, T.C.L., Amato-Lourenço, L.F., da Silva, G.T., Saldiva de André, C.D., de André, P.A., Barrozo, L.V., Singer, J.M., Saldiva, P.H.N., Saiki, M., Locosselli, G.M.: The Use of Tree Barks to Monitor Traffic Related Air Pollution: A Case Study in São Paulo–Brazil. Front. Environ. Sci. 6, 72 (2018).
2. Ejidike, I.P., Onianwa, P.C.: Assessment of Trace Metals Concentration in Tree Barks as Indicator of Atmospheric Pollution within Ibadan City, South-West, Nigeria. J. Anal. Methods Chem. 2015, 243601 (2015).
3. Onder, S., Dursun, S.: Air borne heavy metal pollution of Cedrus libani (A. Rich.) in the city centre of Konya (Turkey). Atmos. Environ. 40, 1122–1133 (2006).
4. Vieira, J., Matos, P., Mexia, T., Silva, P., Lopes, N., Freitas, C., Correia, O., Santos-Reis, M., Branquinho, C., Pinho, P.: Green spaces are not all the same for the provision of air purification and climate regulation services: The case of urban parks. Environ. Res. 160, 306–313 (2018).
5. Conde, A.: Efectos nocivos de la contaminación ambiental sobre la embarazada. Rev. Cuba. Hig. Epidemiol. 51, 226–238 (2013).
6. Baslar, S., Dogan, Y., Bag, H., Elci, A.: Trace element biomonitoring by needles of Pinus brutia Ten. from western Anatolia, Turkey. Fresenius Environ. Bull. 12, 450–453 (2003).
7. Kaya, Gökçe, Yaman, M.: Trace Metal Concentrations in Cupressaceae leaves as biomonitors of environmental pollution. Trace Elem. Electrolytes. 25, 156–164 (2008).
8. Lijteroff, R., Lima, L., Prieri, B.: Uso de líquenes como bioindicadores de contaminación atmosférica en la ciudad de San Luis, Argentina. Rev. Int. Contam. Ambient. 25, 111–120 (2009).
9. Gadzala-Kopciuch, R., Berecka, B., Bartoszewicz, J., Buszewski, B.: Some considerations about bioindicators in environmental monitoring. Pol. J. Environ. Stud. 13, (2004).
10. Karaaslan, N.M., Yaman, M.: Determination of Nickel and Chromium in Pinus Nigra L., Cedrus Libani, and Cupressus Arizonica Leaves to Monitor the Effects of Pollution in Elazig (Turkey). Instrum. Sci. Technol. 41, 335–348 (2013).
11. Önder, S., Dursun, S., Gezgin, S., Demirbaş, A.: Determination of heavy metal pollution in grass and soil of city centre green areas (Konya, Turkey). Pol. J. Environ. Stud. 16, 145–154 (2007).
12. El-Khatib, A.A., Barakat, N.A., Youssef, N.A., Samir, N.A.: Bioaccumulation of heavy metals air pollutants by urban trees. Int. J. Phytoremediation. 22, 210–222 (2020).
13. Chrabąszcz, M., Mróz, L.: Tree Bark, a valuable source of information on air quality. Pol. J. Environ. Stud. 26, 453–466 (2017).
14. Catinon, M., Ayrault, S., Boudouma, O., Asta, J., Tissut, M., Ravanel, P.: Atmospheric element deposit on tree barks : the opposite effects of rain and transpiration. Ecol. Indic. 14, 170 (2012).
15. Guéguen, F., Stille, P., Lahd Geagea, M., Boutin, R.: Atmospheric pollution in an urban environment by tree bark biomonitoring--part I: trace element analysis. Chemosphere. 86, 1013–1019 (2012).
16. Nabulo, G., Oryem Origa, H., Nasinyama, G.W., Cole, D.: Assessment of Zn, Cu, Pb and Ni contamination in wetland soils and plants in the Lake Victoria basin. Int. J. Environ. Sci. Technol. 5, 65–74 (2008).
17. Nawrot, N., Wojciechowska, E., Rezania, S., Walkusz-Miotk, J., Pazdro, K.: The effects of urban vehicle traffic on heavy metal contamination in road sweeping waste and bottom sediments of retention tanks. Sci. Total Environ. 749, 141511 (2020).
18. Wang, W., Feng, L., Zheng, T., Liu, Y.: The sustainability of ecotourism stakeholders in ecologically fragile areas: Implications for cleaner production. J. Clean. Prod. 279, 123606 (2021).

19. Zgłobicki, W., Telecka, M., Skupiński, S.: Assessment of short-term changes in street dust pollution with heavy metals in Lublin (E Poland)-levels, sources and risks. Environ. Sci. Pollut. Res. Int. 26, 35049–35060 (2019).
20. Suárez-Salas, L., Álvarez Tolentino, D., Bendezú, Y., Pomalaya, J.: Caracterización química del material particulado atmosférico del centro urbano de Huancayo, Perú. Rev. Soc. Quím. Perú. 83, 187–199 (2017).
21. De la Cruz Lezama, M.: Concentración de contaminantes del aire generado por las fuentes móviles en la ciudad de Huancayo 2012. Univ. Nac. Cent. Perú. (2015).
22. Xue, W., Jiang, Y., Shang, X., Zou, J.: Characterisation of early responses in lead accumulation and localization of Salix babylonica L. roots. BMC Plant Biol. 20, 296 (2020).
23. Chandra, R., Kang, H.: Mixed heavy metal stress on photosynthesis, transpiration rate, and chlorophyll content in poplar hybrids. For. Sci. Technol. 12, 55–61 (2016).
24. He, J., Ma, C., Ma, Y., Li, H., Kang, J., Liu, T., Polle, A., Peng, C., Luo, Z.-B.: Cadmium tolerance in six poplar species. Environ. Sci. Pollut. Res. 20, 163–174 (2013).
25. Elobeid, M., Göbel, C., Feussner, I., Polle, A.: Cadmium interferes with auxin physiology and lignification in poplar. J. Exp. Bot. 63, 1413–1421 (2012).
26. Laureysens, I., Blust, R., De Temmerman, L., Lemmens, C., Ceulemans, R.: Clonal variation in heavy metal accumulation and biomass production in a poplar coppice culture: I. Seasonal variation in leaf, wood and bark concentrations. Environ. Pollut. Barking Essex 1987. 131, 485–494 (2004).
27. Alcala, J., Ortíz, J., Hernandez, A., Filippini, M., Martínez Carretero, E., Diaz-Flores, P.: Capacity of Two Vegetative Species of Heavy Metal Accumulation. Rev. Fac. Cienc. Agrar. 50, 123–139 (2018).
28. Yarupaitán Galván, G., Albán Castillo, J.: Fanerógamas de la provincia de Huancayo, Perú. Rev. Peru. Biol. 11, 193–202 (2004).
29. Baltazar, H., Baltazar, D., Baltazar, N., Alvarez, D.: Primer registro de Acanthoscelides obtectus (Coleoptera: Chrysomelidae: Bruchinae) en Senna multiglandulosa (Fabaceae) de la región andina central del Perú. Cienc. Tecnol. Agropecu. 22, e2290–e2290 (2021).
30. Franco, W., Peñafiel, M., Cerón, C., Freire, E.: Biodiversidad productiva y asociada en el Valle Interandino Norte del Ecuador. Bioagro. 28, 181–192 (2016).
31. Singh, S., Singh, S.K., Yadav, A.: A Review on Cassia species: Pharmacological, Traditional and Medicinal Aspects in Various Countries. Am. J. Phytomedicine Clin. Ther. 1, 291–312 (2013).
32. Evangelou, M.W.H., Ebel, M., Schaeffer, A.: Chelate assisted phytoextraction of heavy metals from soil. Effect, mechanism, toxicity, and fate of chelating agents. Chemosphere. 68, 989–1003 (2007).
33. Reeves, R.D., Baker, A.J.M., Jaffré, T., Erskine, P.D., Echevarria, G., van der Ent, A.: A global database for plants that hyperaccumulate metal and metalloid trace elements. New Phytol. 218, 407–411 (2018).
34. Udiba, U.U., Antai, E.E., Akpan, E.R.: Assessment of Lead (Pb) Remediation Potential of Senna obtusifolia in Dareta Village, Zamfara, Nigeria. J. Health Pollut. 10, 200301 (2020).
35. U. S. Department of Health & Human: Agency for Toxic Substances and Disease Registry, https://www.atsdr.cdc.gov/index.html, last accessed 2021/11/13.
36. Bokowski, L.V.V., Sobrinho, R.B., Armijo, C.J.V., Dani, C., Henriques, J.A.P., Funchal, C.: Method validation for determination of metals in Vitis labrusca L. grapevine leaf extracts by inductively coupled plasma mass spectrometry (ICP-MS). An. Acad. Bras. Ciênc. 88, 2247–2255 (2016).
37. Cruz, A.R.H.D.L., Ayuque, R.F.O., Cruz, R.W.H.D.L., López-Gonzales, J.L., Gioda, A.: Air quality biomonitoring of trace elements in the metropolitan area of Huancayo, Peru using transplanted Tillandsia capillaris as a biomonitor. An. Acad. Bras. Cienc. 92, e20180813 (2020).
38. INEI: Censos Nacionales 2017: XII de Población, VII de Vivienda y III de Comunidades Indígenas, https://www.inei.gob.pe/media/MenuRecursivo/publicaciones_digitales/Est/Lib1539/libro.pdf, last accessed 2021/04/17.

39. Corilla Huaman, C.P.: Propuesta de mejora del nivel de servicio del tránsito vehicular en la Av. Huancavelica – tramo Av. 13 de noviembre y Paseo La Breña en la ciudad de Huancayo. Univ. Cont. (2018).
40. Garay Flores, R.K.: Análisis de la congestión vehicular y peatonal en la ciudad de Huancayo. Univ. Nac. Ing. (2015).
41. SINIA: Sistema Nacional de Informaciòn Ambiental – MINAM, https://sinia.minam.gob.pe/indicador/966, (2022).
42. Garay Flores, R.K.: Análisis de la congestión vehicular y peatonal en la ciudad de Huancayo. Univ. Nac. Ing. (2015).
43. Hilario Roman, N.: Emisiones contaminantes de vehículos del distrito de Huancayo. Univ. Nac. Cent. Perú. (2017).
44. Machado, A., García, N., García, C., Acosta, L., Córdova, A., Linares, M., Giraldoth, D., Velásquez, H.: Contaminación por metales (Pb, Zn, Ni y Cr) en aire, sedimentos viales y suelo en una zona de alto tráfico vehicular. Rev. Int. Contam. Ambient. 24, 171–182 (2008).
45. Odukoya, O.O., Arowolo, T.A., Bamgbose, O.: Pb, Zn, and Cu levels in tree barks as indicator of atmospheric pollution. Environ. Int. 26, 11–16 (2000).
46. Jeong, H., Ra, K.: Characteristics of Potentially Toxic Elements, Risk Assessments, and Isotopic Compositions (Cu-Zn-Pb) in the PM10 Fraction of Road Dust in Busan, South Korea. Atmosphere. 12, 1229 (2021).
47. Zafra-Mejía, C.A., Luengas-Pinzón, E.C., Temprano-González, J.: Influencia del tráfico en la acumulación de metales pesados sobre vías urbanas: Torrelavega (España)-Soacha (Colombia). Rev. Fac. Ing. Univ. Antioquia. 146–160 (2013).
48. Alcalá, J., Rodríguez Ortíz, J.C., Hernández Montoya, A., Díaz Flores, P.E., Filippini, M.F., Martínez Carretero, E.: Cortezas de Prosopis laevigata (Fabaceae) y Schinus molle (Anacardiaceae) como bioindicadoras de contaminación por metales pesados. Rev. Fac. Cienc. Agrar. (2015).
49. Barbes, L., Barbulescu, A., Radulescu, C., Stihi, C., Chelarescu, E.: Determination of Heavy Metals in Leaves and Bark of Populus Nigra L. by Atomic Absorption Spectrometry. Romanian Rep. Phys. 66, (2014).
50. Sujatha, K.: Determination of Bioaccumulation of Heavy Metals in Leaves, Bark and in Soils by Atomic Absorption Spectrometry. Int. J. Res. Rev. 6, 90–94 (2019).
51. Samara, T., Spanos, I., Platis, P., Papachristou, T.G.: Heavy Metal Retention by Different Forest Species Used for Restoration of Post-Mining Landscapes, N. Greece. Sustainability. 12, 4453 (2020).
52. Ugulu, I., Dogan, Y., Baslar, S., Varol, O.: Biomonitoring of trace element accumulation in plants growing at Murat Mountain. Int. J. Environ. Sci. Technol. 9, 527–534 (2012).
53. Ciftci, H., Er Caliskan, C., Aslanhan, E., Aktoklu, E.: Monitoring of heavy metal pollution by using Populus nigra and Cedrus libani. Sigma J. Eng. Nat. Sci. 39, 367 (2021).
54. Turkyilmaz, A., Cetin, M., Sevik, H., Isinkaralar, K., Saleh, E.A.A.: Variation of heavy metal accumulation in certain landscaping plants due to traffic density. Environ. Dev. Sustain. 22, 2385–2398 (2020).
55. Ouyang, J., Li, B., Li, C., Shang, X., Zou, J.: Cadmium Effects on Mineral Accumulation and Selected Physiological and Biochemical Characters of Salix babylonica L. Pol. J. Environ. Stud. 26, 2667–2676 (2017).
56. FengLiang, Z., WeiDong, Y.: Review on application of willows (Salix spp.) in remediation of contaminated environment. Acta Agric. Zhejiangensis. 29, 300–306 (2017).
57. Sociedad Nacional de Minería, Petróleo y Energía: Decreto Supremo N° 021-2007-EM – SNMPE, https://www.snmpe.org.pe/repositorio-legislacion/272-hidrocarburos/4081-decreto-supremo-n-021-2007-em.html, last accessed 2021/11/16.
58. Tello, V.: Las emisiones de plomo del Complejo Metalúrgico Doe Run y su relación con las infecciones respiratorias agudas de los pobladores de la localidad de Huaynacancha, Distrito de La Oroya – Yauli – Junín – Perú. Rev. Inst. Investig. Fac. Minas Metal. Cienc. Geográficas. 17, (2014).

59. Romero Ledezma, K.P.: Contaminación por metales pesados. Rev. Científica Cienc. Médica. 12, 45–46 (2009).
60. Méndez, J.P., Ramírez, C.A.G., Gutiérrez, A.D.R., García, F.P.: Contaminación y fitotoxicidad en plantas por metales pesados provenientes de suelos y agua. 17 (2009).
61. MINAM, M. del: Decreto Supremo N° 020-2021-MINAM, https://www.gob.pe/institucion/minam/normas-legales/2035448-020-2021-minam, (2021).
62. Castillo, B., Ruiz, J.O., Manrique, M., Pozo, C.: Contaminación por plaguicidas agrícolas en los campos de cultivos en Cañete (Perú). Repos.-Inst. – UNAH. (2020).
63. MINAM: Decreto Supremo N°011-2017-MINAM. 4 (2017).
64. Perelman, P., Faggi, A., Castro, M., Carretero, E.M.: Pollution trends using bark of morus alba in the cities of buenos aires and mendoza (Argentina). Rev. Árvore. 34, 505–511 (2010). https://doi.org/10.1590/S0100-67622010000300014.

Nipa Sap (*Nypa fruitican*) as Supplementary Feedstock for Bioethanol Production in the Philippines: The Pamplona Experience

N. E. R. Mateo, R. A. Ulep, J. I. Rosario, M. P. Lucas, S. C. Agrupis, J. M. Sagadraca, and C. C. Baga

Abstract

The bioethanol demand in the Philippines does not correspond with the available supply for the E10 and E20 requirements of the Philippine Biofuels Act. This paper aims to introduce *Nypa fruitican* as an additional bioethanol feedstock for the country. In Pamplona alone, more than 1000 ha of nipa stands can produce an additional 720,000–1,008,000 liters of bioethanol if 30% of these naturally grown nipa stands are tapped for bioethanol production based on 5–7% alcohol

N. E. R. Mateo (✉)
DME-R, Mapua University, Manila, Philippines

Mariano Marcos State University, Batac, Philippines
e-mail: nrmateo@mmsu.edu.ph

R. A. Ulep
Mariano Marcos State University, Batac, Philippines

Chemistry Department, College of Arts and Sciences, Mariano Marcos State University, Batac, Philippines

J. I. Rosario
Mariano Marcos State University, Batac, Philippines

Forestry Department, College of Agriculture, Forestry, and Sustainable Development, Mariano Marcos State University, Batac, Philippines

M. P. Lucas
Mariano Marcos State University, Batac, Philippines

Department of Economics, College of Business Economics and Accountancy, Mariano Marcos State University, Batac, Philippines

J. T. Anderson (ed.), *Advances in Environmental Sustainability*, Springer Proceedings in Earth and Environmental Sciences,
https://doi.org/10.1007/978-3-031-26365-1_2

yield. In the Philippines, there are 8000 ha of reported nipa stands. In the Pamplona experience, more than 100,000 liters of sap were collected and processed as hydrous nipa bioethanol by four community partners in three different barangays using the Mariano Marcos State University village-scale reflux distillation facility and technologies. The hydrous ethanol produced in Pamplona was dehydrated in the distillation facility of Far East Alcohol Corporation on August 22, 2019, and marks the first nipa bioethanol distillation in the Philippines, suggesting that nipa is a technically viable feedstock despite the challenges encountered during this experience.

Keywords

Nipa Sap · Hydrous bioethanol production · Pamplona experience

1 Introduction

Fifty-seven percent of the total bioethanol requirement in the Philippines in 2015 alone was imported from the United States [1]. In 2020, the Department of Energy in the Philippines projected that 4302 million liters of gasoline and 860 million liters of bioethanol are expected to increase annually by 1.9% and 9.7% for gas and bioethanol, respectively, and the reported production capacity of 12 bioethanol plants in the country is only 350.5 million liters per year [2]. Assuming that these bioethanol plants are operating within their declared capacity, there is still a supply gap of about 59%. In addition to the distillation facility requirements of the country, a sustainable bioethanol feedstock must also be considered for the country's bioethanol requirements. Sugarcane, corn, cassava, and sweet sorghum were identified as potential bioethanol production feedstock [3, 4]. However, devoting sugarcane as feedstock may increase the prices of domestic products. Sugar is one of the major input raw materials in the food and beverage industries. Increasing mass production for crops such as sugarcane, sweet sorghum, corn, and cassava leads to farmland conversion for bioethanol rather than food.

This paper aims to introduce other potential feedstocks, such as the Nipa palm (*Nypa fruitican*) [5, 6], that prefer to grow naturally in a non-farmland area similar to mangroves [7] and may help ensure consistent and sustainable supply in large-scale ethanol distilleries of the country.

S. C. Agrupis
Mariano Marcos State University, Batac, Philippines

Biology Department, College of Arts and Sciences, Mariano Marcos State University, Batac, Philippines

J. M. Sagadraca · C. C. Baga
National Bioenergy Research and Innovation Center, Paoay, Philippines

2 The *Nypa fruitican*

Nipa palm had a pantropical distribution 13–63 million years ago, as suggested by the Eocene and Miocene fossils found in Europe, the Middle East, and North America, and the Paleocene strata in Brazil [8]. Nipa is a palm species commonly found in the mangroves of tropical regions and been tapped to produce fresh juice, fermented drinks, syrup, and raw and refined sugar some 50–100 years ago [9]. Unlike other feedstock such as sugarcane and corn that needs to be planted, cultivated, and maintained using fertilizers, the nipa palms grow naturally and invade open mangroves. The nipa sap is extracted from the nipa palm's peduncle.

In the Philippines, Mariano Marcos State University, thru its National Bioenergy Research and Innovation Center (MMSU-NBERIC), used nipa sap in its research on bioethanol production as a blended and unblended fuel, and in its ethanol production to address the scarcity of alcohol in its community in this time of the pandemic.

2.1 Some Sitting and Occurrences in the Philippines

The most extensive nipa stands are found in Indonesia, Papua New Guinea, and the Philippines, with 8000 hectares [8]. These are commonly observed in the Philippines thru its mangroves but thrive in brackish waters and are used as construction materials (either as roof or walls) for the famous "bahay kubo," also known as "nipa hut"; vinegar; and wine as observed in Cagayan, and in Pangasinan. In Pangasinan, their nipa vinegar is famous as "sukang paombong." In Northern Luzon, vast nipa stands were present in Cagayan, including municipalities of Sanches, Pamplona, Abulog, Aparri, and Tuguegarao. Nipa stands are also present in other parts of Luzon, such as in Bugalion, Pangasinan, Dagupan, Mulanay, Quezon, Palawan, and Marinduque.

Studies of nipa sampling sites conducted by Mantiquillia et al. [10] also mentioned areas of Mindanao such as Pagadian, Tudela, Linamon, Opol, Kabadbaran, Barobo, Tagum, Talomo, and Hagonaoy, including the study conducted by Carandang et al. [11] where the occurrence of nipa is present in Bohol. In addition to Bohol, the presence of nipa palm in the Visayas is also present in Antique and Aklan. Cui et al. [6] also claimed in their study that Northern Samar has abundant nipa stands with significant potential as sources of bioethanol feedstock.

Comparison to Other Bioethanol Feedstock in the Philippines Unlike other crops cultivated and maintained thru agricultural fertilizers, nipa palms grow naturally in the brackish waters as mangroves. Its leaves are used as construction materials in rural areas as roofs or walls of their houses, while the sap is extracted and processed as vinegar or wine [11, 12]. This potential feedstock does not require farmland to be cultivated and, thus, does not compete in the food versus fuel issue.

Bioethanol Production from Nipa The number of extractable nipa peduncle or stalks per hectare varies depending on the type and location of nipa stands, as reflected in the study conducted by Hidayat [13]. Likewise, the bioethanol production and sap yield range from 20.45 to 136.65 liters of bioethanol per hectare per day based on 0.5 to 2.0-liter of sap that can be extracted from nipa palm [17]. The ethanol yield is 8.95–15% [13]. The yield and chemical composition of nipa may depend on plant variety, maturity, and type of plantation [14]. Jaichuedee et al. (2016) also mentioned that good management practice must be established in the nipa stands.

3 The Pamplona Experience

3.1 Area and Location of Nipa Stands

The study was conducted in Pamplona, Cagayan, located on the northern tip of the Philippines. It was found that 2214 hectares of nipa stands are in Abulug and Pamplona. About 1200 hectares of nipa stands are located in Pamplona. Information on nipa occurrences in Cagayan was determined from secondary sources, including government agencies such as the Department of Environment and Natural Resources of Cagayan (provincial and municipal), the Department of Agriculture, and the Provincial Planning and Development Office.

Density of Nipa The assessment of the density of nipa stands was identified by the municipal mayor and the municipal agriculture office of Pamplona. Both recommended Barangay Cabaggan in Pamplona, Cagayan, with 100 hectares of nipa stands and their local wine production using nipa sap. The nipa stands were inventoried using randomly selected 20 m × 20 m sample plots. Separate sample plots were established in three sites: intensively managed stands, unmanaged/minimally managed stands, and young stands (less than 5 years old).

The estimated density of nipa in Barangay Cabaggan, Pamplona, Cagayan was highest in minimally managed stands (3,457 plants ha-1), followed by the young stands (3,212 plants ha-1). The intensively managed stands had the lowest density (2787 plants ha^{-1}). However, this is relatively high based on Lestari et al. [15] of 861–913 stalks per hectare, but close to the 1526–3550 trees per hectare in the study conducted by Roberson et al. [16].

Sap Collection and Sap Yield The extracted sap varies with the nipa tapping season. It appeared from field interviews of the sap collectors, also known as "kappatero" or "mag tatata," that sap collection from one peduncle ranges from 1 liter to 2.5 liters using bamboo vessels, also known as "tokil." The sap yield, when reconciled to the reported sap yield of 0.2–2 liters [17] in the study conducted by Kurniawan et al. (2018) and to the study conducted by Garba et al. (2019) for a maximum sap yield of 1.0, 1.7, and 2.5 liters per day per palm for the low, medium,

Table 1 Sap collection using different techniques in Barangay Cabaggan

Technique	Volume of Sap @14–16 initial Brix, mL		Occurrences of >600 mL collection, %
	Min	Max	
Aerobic-Aerobic	450	1150	60
Aerobic-Anaerobic	500	1130	80
Anaerobic-Anaerobic	1030	1340	50

and high yield scenarios, respectively [18] in Nigeria appeared to have similarities based on the villager's interviews in Barangay Cabaggan. The initial report of the study paved the way for the first hydrous bioethanol production in Pamplona from 2018 to 2019. However, the sap extraction conducted at the Barangay Cabaggan site in 2019 was validated using three different sap collection techniques (Table 1).

Ten random samples are distributed in two 20 × 20 plots, ten samples for each method. The three methods were conducted in anticipation of the alcohol yield determination and, simultaneously, to introduce a better sap quality collection. The anaerobic sap collection prevents dirt and insect contamination. Aerobic-aerobic technique means aerobic sap collection in the peduncle and aerobic sap collection storage using a developed apparatus similar to the study conducted by Madigal et al. (2020) in developing a closed sap collection vessel [19]. Two separate study locations in Barangay Cabaggan deployed MMSU hydrous bioethanol technology to improve sap collection and yield. Two other barangays (Tabba and Sta. Cruz) of Pamplona also participated in hydrous ethanol production using nipa sap.

3.2 The Hydrous Bioethanol Production Run in Pamplona

Four locations in three barangays of Pamplona participated in the research project funded by the Department of Energy, USAID-STRIDE, and the Sugar Regulatory Administration. Table 2 reflects the bioethanol production of the four commissioned recipients in terms of nipa hydrous bioethanol. More than 100,000 liters of nipa sap were collected and distilled using the MMSU 850L batch capacity reflux distillers [20] installed in Pamplona. The alcohol yield varies depending on how strictly the recipients followed the protocols from sap collection up to distillation. The four recipients did not start the production at the same time, but the 7% alcohol yield of Arinabo implies a better implementation of the bioethanol production process.

The First Dehydration Trial of Bioethanol Feedstock Using *Nipa fruitican* The produced hydrous bioethanol from nipa at the village level was further purified at the Far East Alcohol Corporation (FEAC) on August 22, 2019, in Barangay Apalit, Pampanga, and marks the first-ever bioethanol production of anhydrous ethanol using *Nipa frutican* as claimed by the Ethanol Producers Association of the Philippines (EPAP). The hydrous bioethanol at 92.5% average purity enters the

Table 2 Hydrous bioethanol production run in Pamplona

Recipient	Total Sap processed, liters	Equivalent volume of ethanol, 95% grade	Alcohol yield, %v/v
A.B. Eduarte Distilery	20,690	827.6	4.0
Brigido Arinabo	10,010 (+800 L 20% v/v)	774	7.0
Cabaggan Women's Organization	20,800	832	4.0 4.0
Sta. Cruz Multi- Purpose Cooperative	49,399	1678.75	3.4
Total production	**100,899**	**4111**	**4.0**

Table 3 Nipa bioethanol test analysis

Properties and requirements as per Philippine National Standard (PNS)		Test Results
Appearance	Clear and Bright, visibility free from suspended or precipitated contaminants	Passed
Color	Colorless, water-like	Passed
Acidity/Alkalinity, pHe	6.5–9.	7.7
Ethanol Content, %v/v (min)	99.3	99.73
Density @ 20 deg. Celsius, g/cm^3 (max)	0.7915	0.7906
Water Content, % v/v (max)	0.5	0.35
Remarks		**Ok**

pipeline of the FEAC facility with observed regular operation. Table 3, as shown below, was the result of the nipa bioethanol analysis conducted by FEAC during the dehydration run and implies that nipa bioethanol is comparable to other bioethanol feedstock as per bioethanol Philippine National Standard.

Cost of Nipa Hydros Ethanol in the Village Level of Pamplona The cost of hydrous bioethanol from nipa ranges from Php 39.93 to Php 56.90 based on a 7% alcohol yield and nipa buying price per liter of Php 1.25 to Php 2.50 per liter, as reflected in Table 4.

These data are based on the MMSU hydrous distillation facilities deployed in Pamplona, which include feedstock cost (varies from Php 1.25 to Php 2.50), cost of electricity (Php 104), fuel wood (Php 300), maintenance (Php 150), labor (Php 400), and, administrative/profit cost (Php 400).

Cost of Anhydrous Nipa Bioethanol Additional ethanol purification cost is required to dehydrate the hydrous ethanol produced at the village level. This includes the price of steam and electricity and transportation costs of Php 1.18 per liter and Php 2.17 per liter, respectively, based on the FEAC facility where the actual

Table 4 Cost of Nipa Hydrous bioethanol

Feedstock cost per liter, Php	Total production cost, Php	Cost per liter of Hydrous bioethanol, Php		
		4%	6%	7%
1.25	2354	69.88	46.59	39.93
1.5	2554	75.82	50.55	43.33
1.75	2754	81.76	54.51	46.72
2.0	2954	87.7	58.46	50.11
2.25	3154	93.63	62.42	53.51
2.5	3354	99.57	66.38	56.9

Table 5 Production cost of anhydrous ethanol from Nipa Sap

Cost of hydrous ethanol at different alcohol % yield, Php/liter	Cost of anhydrous ethanol, Php/liter
39.93–56.9 (7%)	43.28–60.25
46.59–66.38 (6%)	49.94–69.73
69.88–99.57 (4%)	73.23–102.92

Table 6 Ethanol production in Pamplona in a year

Nipa stand, hectare	extractable stands, hectare^{-1}	50 days collection, liters	Ethanol production, liters
800	360	14,400,000	720,000–1,008,000

dehydration trial was conducted. To be competitive, the total anhydrous cost per liter using nipa must be maintained at a 7% alcohol yield (Table 5).

Potential as Emerging Feedstock It was claimed by the "kappateros" of Pamplona that there are three nipa sap collecting seasons and 45 days of extraction time for every peduncle of the nipa palm in Pamplona. This is almost half of the 100 days [18], Garba et al. (2019) claimed. However, it appears that the 45 days earlier claimed by the sap collectors was not adequately established due to the erratic weather condition of the sites. However, sap collection season was observed all year round, with peaks in June-July, October-November, and January-February. It was also observed that rains and typhoons causing flooding in the area dictate the sap extraction season for nipa. The potential of nipa bioethanol as bioethanol feedstock is reflected in Table 6. Based on the observed production run, 800 liters of sap can be collected for every hectare of the nipa stand. If 30% of the 1200 hectares of nipa plantation in Pamplona is utilized for bioethanol production and using a conservative 50 days of sap collection (since the average distillation run per location was 20 days per month operating full-time over 6 months or 120 days) using a 5–7% alcohol yield is about 720,000–1,008,000 liters of nipa bioethanol.

4 Challenges of Nipa Bioethanol in Pamplona

4.1 Alcohol Yield

Maintaining the alcohol yield of 7% v/v recovery or greater is very important since the glucose of nipa ranges from 3.4–18.8% [21] based on the study of Thorasin et al. (2018) or 8.98% v/v alcohol content according to Tamunaidu (2013). Only one of the four participating technology recipients strictly followed the anaerobic collection or storage protocols. A high ethanol yield was observed in the Arinabo facility with +7% v/v alcohol yield as the storage of sap collected was held closed, preventing air from infiltrating the sap to cause aerobic fermentation. Following the protocols to maintain an acceptable alcohol yield thru collection and fermentation is a massive challenge on the village scale. The distillers' distillation efficiency also contributed to the overall alcohol yield at 80–90% efficiency.

4.2 Transition of the Traditional Practice to Scientific Method and Reduced Numbers of "Kappateros"

Implementing actual production of nipa bioethanol was quite challenging, especially in the transfer of technology. The community recipients often resist the anaerobic storage of collected nipa, including recycled Pet bottle use. The "kappateros" point out some traditional and cultural beliefs such as collection is only done in the early dawn, setting up of their "tokil" with the presence of another person including birds means misfortune to the person or its family members, etc. All this belief was proven irrelevant in the field, as saps can be collected any time of the day, including witnessing the actual traditional collection processes. One of the reasons for the vast area of minimally managed nipa stands is the decreasing number of incoming generations of "kappateros"; the community already has less interest in the nipa palm industry.

4.3 Erratic Weather in Pamplona

More often, the unpredictable rainy condition in Pamplona is a problem in bioethanol production. This disrupts schedules and causes additional burdens in the preparation, collection, and distillation. Typhoons and flooding in the area destroy the collection set-up and require more than a month to recover and rehabilitate the nipa stands, including the distillation facilities affected by the rainy or stormy conditions.

4.4 Cost of Nipa Sap

The cost of nipa sap for local wine in Pamplona affects the price of nipa sap for bioethanol production. We observed that during sap collection for wine, the cost of sap for the nipa bioethanol distillation increases up to Php 5.00 per liter of sap

5 Conclusion

An estimated 800 liters of sap can be collected from every hectare of nipa stand. If 30% of the 1200 hectares of nipa plantation in Pamplona is utilized for bioethanol production, the potential ethanol that can be produced using 5–7% alcohol yield is 720,000–1,008,000 liters of nipa bioethanol.

For an average price index of Php 61.0 per liter of ethanol from January to March 2020, as reflected in the report of Gatdula et al. (2021), a profit of Php 0.75 to Php 17.72 per liter is realized if the alcohol yield is at least 7% and the price ranges from Php 1.25 to Php 2.50 per liter of sap.

Despite the challenges encountered, the Pamplona experience results show that nipa sap is an alternative and complementary feedstock for anhydrous ethanol production. Producing hydrous ethanol from nipa at the village level has opened more community opportunities, especially for the Nipa farmers. With the continued support of the national and local governments, research on establishing a village-scale dehydration plant in the area must be conducted since the transportation cost of the hydrous nipa bioethanol highly contributed to the total cost of the production of nipa anhydrous bioethanol.

Acknowledgment Special thanks to the following who supported the Nipa Bioethanol Pamplona experience: The late President Romulo Kheyeng of FEAC, Executive Director Ms. Queenie Rojo of EPAP, Engr. Rosemarie Gumera of Sugar Regulatory Administration, the late Pamplona Mayor Aaron Sampaga, the municipality of Pamplona, Senator Sherwin Gatchalian, MMSU-NBERIC staff, Dr. Fiorello Abenes of DOST Balik-scientist, and our Pamplona community recipients: Crisanta Leaño, Quirino Muñoz, Alain Eduarte, and Brigido Arinabo, and to Dr. Aldrin Calderon and Dr. March Manuel of Mapua University.

References

1. J. Beckman and G. Nigatu, "A Report from the Economic Research Service Global Ethanol Mandates: Opportunities for U.S. Exports of Ethanol and DDGS," *Rep. from Econ. Res. Serv. (USDA ERS)*, vol. BIO-05, no. October 2017, pp. 1–32, 2017.
2. K. M. Gatdula, R. B. Demafelis, and B. G. Bataller, "Comparative Analysis of Bioethanol Production from Different Potential Biomass Sources in the Philippines," *Bioethanol Technol.*, Feb. 2021
3. Z. A. Kapasi, A. R. Nair, S. Sonawane, and S. K. Satpute, "Biofuel -An alternative source of energy for present and future," vol. 13, no. 0971, pp. 105–108, 2010.
4. S. Marx, "Cassava as Feedstock for Ethanol Production: A Global Perspective," *Bioethanol Prod. from Food Crop.*, pp. 101–113, Jan. 2019.
5. P. Tamunaidu, N. Matsui, Y. Okimori, and S. Saka, "Nipa (Nypa fruticans) sap as a potential feedstock for ethanol production," *Biomass and Bioenergy*, vol. 52, pp. 96–102, May 2013.
6. K. M. R. C.- Lim *et al.*, "Extraction of Ethanol from Nypa fruticans (Nipa) Palm Fruit," *Asian J. Phys. Chem. Sci.*, vol. 8, no. 4, pp. 41–45, Dec. 2020.
7. S. Eddy and M. Basyuni, "Short Communication: The phenomenon of nipah (Nypa fruticans) invasion in the Air Telang Protected Forest, Banyuasin District, South Sumatra, Indonesia," *Biodiversitas J. Biol. Divers.*, vol. 21, no. 11, pp. 5114–5118, Oct. 2020.
8. "Nypa fruticans (nipa palm)," *2021 Copyright CAB International*, 2019. https://www.cabi.org/isc/datasheet/36772

9. N. Matsui, Y. Okimori, F. Takahashi, K. Matsumura, and N. Bamroongrugsa, "Nipa (Nypa fruticans Wurmb) Sap Collection in Southern Thailand I. Sap Production and Farm Management," *Environ. Nat. Resour. Res.*, vol. 4, p75, 2014.
10. Junaldo A. Mantiquilla, Sandy Faye D. Salmasan, Mariah Krysna A. Obelidhon, Reynaldo G. Abad, "Nutrient Status of Nipa (Nypa fruticans Wurmb.) in Selected Areas of Mindanao, the Philippines | BANWA Series B," *BANWA B (2019) 14: art012*, 2020.
11. M. G. Carandang *et al.*, "Sustainable thatching materials production from nipa (nypa fruticans) in Bohol, Philippines," https://doi.org/10.1080/21580103.2008.9656343, vol. 5, no. 1, pp. 17–22, 2010.
12. P. Saithong, S. Nitipan, and J. Permpool, "Optimization of vinegar production from nipa (Nypa fruticans Wurmb.) sap using surface culture fermentation process," *Appl. Food Biotechnol.*, vol. 6, no. 3, pp. 193–200, 2019.
13. I. W. Hidayat, "Economic Valuation of Nipa Palm (Nypa fruticans Wurmb.) Sap as Bioethanol Material," *IOP Conf. Ser. Earth Environ. Sci.*, vol. 166, no. 1, 2018.
14. J. Jaichuedee, R. Longalee, and C. Musikavong, "Water deprivation as an indicator for evaluating the potential areas of nipa (Nypa fruticans) sap ethanol in Thailand," *J. Clean. Prod.*, 2016.
15. N. S. Lestari and R. F. Noor'an, "Population Density and Habitat Characteristics of Nipa Fruticans in Degraded Mangrove Ecosystem (Case Study in Mahakam Delta, East Kalimantan),"*J. Wetl. Environ. Manag.*, vol. 7, no. 1, pp. 50–59.
16. A. I. Robertson, P. Dixon, P. A. Daniel, and I. Zagorskis, "Primary production in forests of the mangrove palm Nypa fruticans," *Aquat. Bot.*, vol. 167
17. T. Kurniawan, J. Jayanudin, I. Kustiningsih, and M. Adha Firdaus, "Palm Sap Sources,Characteristics, and Utilization in Indonesia," *J. Food Nutr. Res.*, vol. 6, no. 9, pp. 590–596
18. N. A. Garba and S. Abdullahi, "Nipa (Nypa fruticans) Palm: The New 'Green Oil' in Nigeria,"*Int. J. Sci. Glob. Sustain.*, vol. 5, no. 2, 2019.
19. J. P. T. Madigal, B. G. Calixto, M. Duldulao, T. D. Ubiña, and S. C. Agrupis, "Development of Nipa (Nypa fruticans) SAp Closed Collection Vessel," *Agro Bali Agric. J.*, vol. 3, no. 2, pp. 108– 117, 2020.
20. N. E. R. Mateo, S. C. Agrupis, and R. A. Ulep, "An Apparatus for Reflux Distillation Hydrous Ethanol.pdf," *UM Certification 2–201700028*, 2018.
21. T. Thorasin, "A Comparative Study of the Physicochemical, Nutritional Characteristics and Microbiological Contamination of Fresh Nipa Palm (Nypa fruticans) Sap," *Burapha Sci. J. (วารสารวิทยาศาสตร์บูรพา)*, vol. 23, no. 3, pp. 1301–1316, Sep. 2018.

Optimization of Fuel Cell Power Systems Using Biogas from Palm Oil Mill Plant

Woranee Mungkalasiri and Jitti Mungkalasiri

Abstract

Nowadays, biogas is a globally significant alternative source of green energy. In Thailand, agricultural industries generate a vast volume of sewage, which can be utilized to produce biogas from anaerobic fermentation. Due to advanced research and technologies, biogas can produce a more valued option as hydrogen through a tri-reforming process. This hydrogen is introduced to generate power by integrating the tri-reforming process with a fuel cell. Therefore, power generation by a tri-reforming process combined with two types of fuel cells: molten carbonate fuel cell (MCFC) and solid oxide fuel cell (SOFC), was studied in this work by the Aspen Plus simulation program. Biogas obtained from a palm oil mill plant was employed as feedstock. The effect of several parameters, including tri-reforming temperature, S/C molar ratio, O_2/C molar ratio, MCFC temperature, and SOFC temperature on power generation at atmospheric pressure, was investigated to find the optimal operating conditions maximizing hydrogen and power. The simulation results showed that the optimal operating conditions of tri-reforming were 700 °C of tri-reforming temperature, 3.0 S/C molar ratio, and 0.1 O_2/C molar ratio. In addition, the optimal temperatures of MCFC and SOFC were 750 °C and 1000 °C, respectively. Moreover, the system efficiency was evaluated, and the power generation system with molten carbonate fuel cell produced 2809 W/m^2 of power density with 63.27% of system efficiency. The

W. Mungkalasiri (✉)
Chemical Engineering Department, Engineering Faculty, Thammasat School of Engineering, Thammasat University, Pathumthani, Thailand
e-mail: pworanee@engr.tu.ac.th

J. Mungkalasiri
Technology and Informatics Institute for Sustainability, National Science and Technology Development Agency, Pathumthani, Thailand

J. T. Anderson (ed.), *Advances in Environmental Sustainability*, Springer Proceedings in Earth and Environmental Sciences,
https://doi.org/10.1007/978-3-031-26365-1_3

power generation system with a solid oxide fuel cell made 5526 W/m^2 of power density with 76.09% of system efficiency.

Keywords

Biogas · Molten carbonate fuel cell · Process simulation · Solid oxide fuel cell

1 Introduction

Biogas is an important source of renewable energy in Thailand. Being an agricultural-based country, various types of raw materials, such as waste or wastewater from agro-industries, can be used for biogas production [1]. In palm oil mill plants, large quantities of water are used; consequently, large amounts of wastewater are also generated. The major steps in producing wastewater are steaming and extracting crude palm oil. The wastewater produced contains many organic compounds, which are suitable as an alternative resource for making renewable energy in the form of biogas. Moreover, biogas can be used for hydrogen production.

Currently, the largest sources of power generation are fossil fuels, which are non-renewable, unsustainable, and pollute the environment. Therefore, renewable energy sources have been developed to replace the usage of fossil fuels. Fuel cells represent one alternative energy source. They are electrochemical devices that directly convert chemical energy from fuels into electrical energy without combustion. The fuel-cell types that operate co-generated heat and electricity are molten carbonate and solid oxide fuel cells. They yield high efficiency and lower emissions than any other types of fuel cells [2].

Using coal as a raw material to produce electricity will produce carbon dioxide contributing to global warming. In addition, coal is a non-renewable energy source. Based on various research [3, 4] on power generation by fuel cells, researchers found that molten carbonate fuel cells (MCFC) and solid oxide fuel cells (SOFC) are highly efficient in generating electrical energy. Therefore, this research studied biogas from a palm oil mill plant for power production via a tri-reforming process integrated with two types of fuel cells: molten carbonate and solid oxide fuel cells. The Aspen Plus simulation program modeled the system.

2 Methodology

2.1 Simulation of Tri-Reforming Process

Biogas is fed to a tri-reforming, producing hydrogen. The tri-reforming process consists of dry reforming (DR), steam reforming (SR), and partial oxidation (POX). The water-gas shift reaction (WGS) also occurs [5]. The reactions are

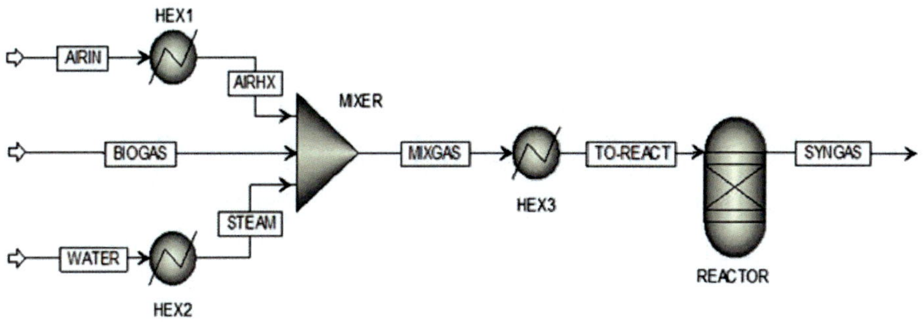

Fig. 1 Simulation block diagram of the tri-reforming process

$$\mathrm{DR: CH_4 + CO_2 \leftrightarrow 2CO + 2H_2} \qquad \Delta H = 247.3\mathrm{kJ/mol}$$
$$\mathrm{SR: CH_4 + H_2O \leftrightarrow CO + 3H_2} \qquad \Delta H = 206.3\mathrm{kJ/mol}$$
$$\mathrm{POX: CH_4 + 1/2O_2 \leftrightarrow CO + 2H_2} \qquad \Delta H = -35.6\mathrm{kJ/mol}$$
$$\mathrm{WGS: CO + H_2O \leftrightarrow CO_2 + H_2} \qquad \Delta H = -41.9\mathrm{kJ/mol}$$

We used the Aspen Plus process design and simulation program to study the tri-reforming process (Fig. 1). Soave-Redlich-Kwong (SRK) Equation of State (EOS) was chosen [5]. Biogas from a palm oil mill plant was used, which consisted of 71% of methane (CH_4) and 29% of carbon dioxide (CO_2) with a flow rate of 33 kmol/h [6]. Biogas was then mixed with air and water. After that, the mixture was heated and fed through a reactor (RGibbs reactor model), which caused the tri-reforming reaction.

In this study, the tri-reforming process was operated at atmospheric pressure because, at high pressure, the production of hydrogen (H_2) would be decreased, and the accumulation of carbon formation would occur. Using the Aspen Plus process simulation, we validated the model to verify the correctness of the tri-reforming process [5].

2.2 Simulation of Fuel Cells

A fuel cell is an electrochemical energy conversion device that directly converts one part of chemical energy into electrical energy by consuming hydrogen-rich fuel and oxidant. Since many fuel cell types exist, we studied high-temperature fuel cells: the molten carbonate fuel cell (MCFC) and solid oxide fuel cell (SOFC). Because high-temperature fuel cells generally have the great advantage that they have the potential to be independent of pure hydrogen [2].

The Aspen Plus program was used for the simulation of the molten carbonate fuel cell (MCFC) and solid oxide fuel cell (SOFC) (Fig. 2). In this study, we validated fuel cell models (MCFC and SOFC) with the literature [7, 8].

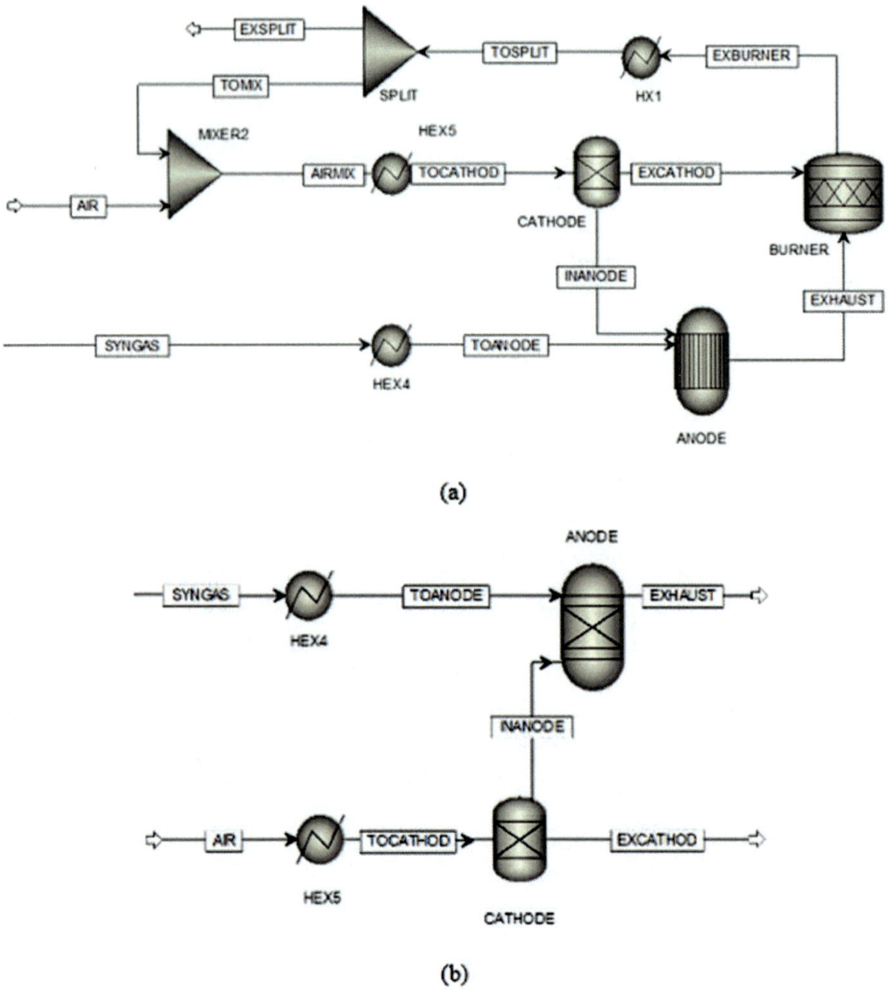

Fig. 2 Simulation block diagram of fuel cells (**a**) MCFC; (**b**) SOFC

2.3 System Performance Evaluation

The MCFC/SOFC fuel cell performances were evaluated by power density output. The power density was calculated via the electrochemical model and input parameters for the MCFC model [7] and the SOFC model [8]. Moreover, the integrated system performances were determined in terms of system efficiency [9].

$$System\ \ efficiency = \frac{P_{MCFC/SOFC} + Q_{out}}{x_{CH_4} LHV_{CH_4} F_{Biogas} + Q_{in}} \times 100\% \quad (1)$$

where F_{Biogas} was the feed flow rate of biogas
LHV_{CH4} was lower heating value of methane

P_{MCFC} was power production from MCFC
P_{SOFC} was power production from SOFC
Q_{in} was heat input
Q_{out} was heat output
x_{CH4} was mole fraction of methane

3 Results and Discussion

We evaluated the optimal operating conditions of an integrated system of biogas tri-reforming and fuel cells (MCFC and SOFC). The effect of important operating variables including tri-reforming temperature (200–1200 °C), steam to carbon (S/C) molar ratio (0.1–3.0), oxygen to carbon (O_2/C) molar ratio (0.1–1.0), MCFC temperature (600–750 °C) and SOFC temperature (800–1000 °C) on power generation at atmospheric pressure was studied.

3.1 Tri-Reforming Process

We evaluated the influences of temperature and steam to carbon ratio on a mole basis (S/C), which affected the flow rate of hydrogen (H_2) produced by the tri-reforming process. We studied the temperature range of 200–1200 °C, steam to carbon ratio (S/C) of 0.1–3.0 at a pressure of 1 atm, and oxygen to carbon ratio (O_2/C) of 0.1. We produced higher hydrogen gas by increasing temperature and steam to carbon ratio (S/C) (Fig. 3a). Because tri-reforming reactions react well at high temperatures, when the temperature increases, it causes a higher forward reaction, leading to higher hydrogen gas produced. However, if the temperature increased above 700 °C, hydrogen gas decreased due to the reverse water-gas shift reaction [5]. Thus, maximum hydrogen yield was achieved at the temperature of 700 °C and the steam to carbon ratio (S/C) of 3.0.

The influence of temperature and oxygen to carbon ratio (O_2/C), which affected the flow rate of hydrogen (H_2) produced by the tri-reforming process, was investigated. We studied the influences of 200–1200 °C temperatures and oxygen to carbon ratios (O_2/C) of 0.1–1.0 at a pressure of 1 atm and steam to carbon ratio (S/C) of 3.0. According to the results in Fig. 3b, less hydrogen gas was produced by increasing the oxygen to carbon ratio. Because the excess amount of oxygen caused complete combustion instead of partial oxidation, this led to less hydrogen gas produced from partial oxidation. Hence, the maximum hydrogen yield was achieved at the temperature of 700 °C and the oxygen to carbon ratio of 0.1. However, the oxygen to carbon ratio should not be less than 0.1 because, with less oxygen, it was unsuitable for the partial oxidation reaction.

The influence of steam to carbon ratio (S/C) and oxygen to carbon ratio (O_2/C), which affected the flow rate of hydrogen (H_2) produced by the tri-reforming process, was evaluated. We studied the S/C ratio and O_2/C ratio with a range between

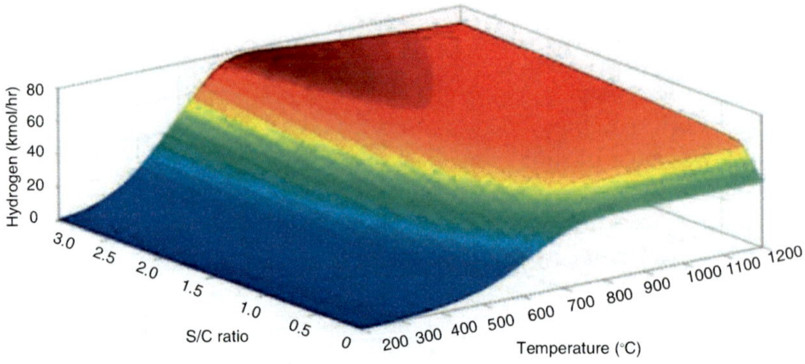

(a)

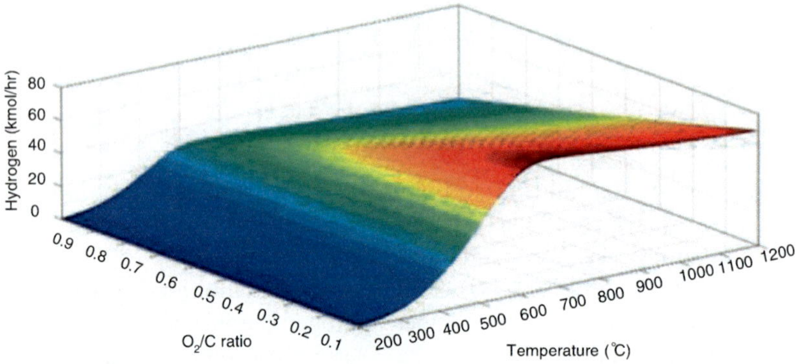

(b)

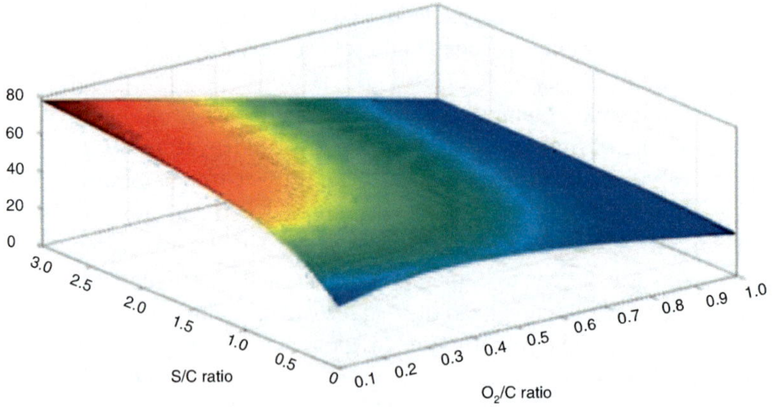

(c)

Fig. 3 The effect of (**a**) temperature and steam to carbon ratio (S/C); (**b**) temperature and oxygen to carbon ratio (O_2/C); (**c**) steam to carbon ratio (S/C) and oxygen to carbon ratio (O_2/C) on H_2 production from biogas tri-reforming process

0.1–3.0 and 0.1–1.0, respectively, at a pressure of 1 atm and temperature of 700 °C (Fig. 3c). We found more hydrogen gas was produced by increasing the steam to carbon ratio and decreasing the oxygen to carbon ratio because the steam reforming reaction and water-gas shift reaction occurred.

Under optimal operating conditions; 700 °C tri-reforming temperature, 3.0 S/C molar ratio, and 0.1 O_2/C molar ratio, the maximum hydrogen production was 73.79 kmol/h.

3.2 Fuel Cells (MCFC and SOFC)

The influence of MCFC temperature (600–750 °C) and SOFC temperature (800–1000 °C) that operated at atmospheric pressure (1 atm) with maximum hydrogen from the tri-reforming process was studied. The current density was specified at 5500 A/m^2 for MCFC and 4500 A/m^2 for SOFC. Increasing temperature caused higher voltage and power density (Fig. 4). Increasing temperature resulted in a more efficient electrochemical reaction and reduced the voltage loss [4]. However, higher temperature operation resulted in the short life of fuel cells due to faster corrosion inside the cells. Therefore, the optimal operating temperature for molten carbonate fuel cells was 750 °C and for solid oxide fuel cells was 1000 °C generating a maximum power density of 2809 W/m^2 and 5526 W/m^2, respectively.

In addition, the performances of integrated systems of biogas tri-reforming and fuel cells (MCFC and SOFC) were measured by system efficiency evaluation. The results were found that using biogas from palm oil mill plant for power production via the tri-reforming process with molten carbonate fuel cell (MCFC) had a system efficiency of 63.27%. For power production operated by the tri-reforming process with solid oxide fuel cell (SOFC), it had a system efficiency of 76.09%.

4 Conclusion

The Aspen Plus program evaluated the simulation study of integrated biogas tri-reforming with MCFC/SOFC systems. We used biogas from a palm oil mill plant (33 kmol/h) as feedstock for power generation. The effects of important operating variables were studied: tri-reforming temperature, S/C molar ratio, O_2/C molar ratio, MCFC temperature, and SOFC temperature. This study evaluated the optimal operating conditions of the integrated systems to produce maximum hydrogen from the tri-reforming process and power generation from fuel cells. The results revealed that the optimal operating conditions of the tri-reforming process provided the maximum hydrogen production (73.79 kmol/h) were 700 °C of tri-reforming temperature, 3.0 S/C molar ratio, and 0.1 O_2/C molar ratio.

Moreover, the optimal operating conditions to generate maximum power were 750 °C of MCFC temperature, which produced 2809 W/m^2 of power density, and 1000 °C of SOFC temperature, which had 5526 W/m^2 of power density. In addition, the performances of integrated systems were evaluated by system efficiency

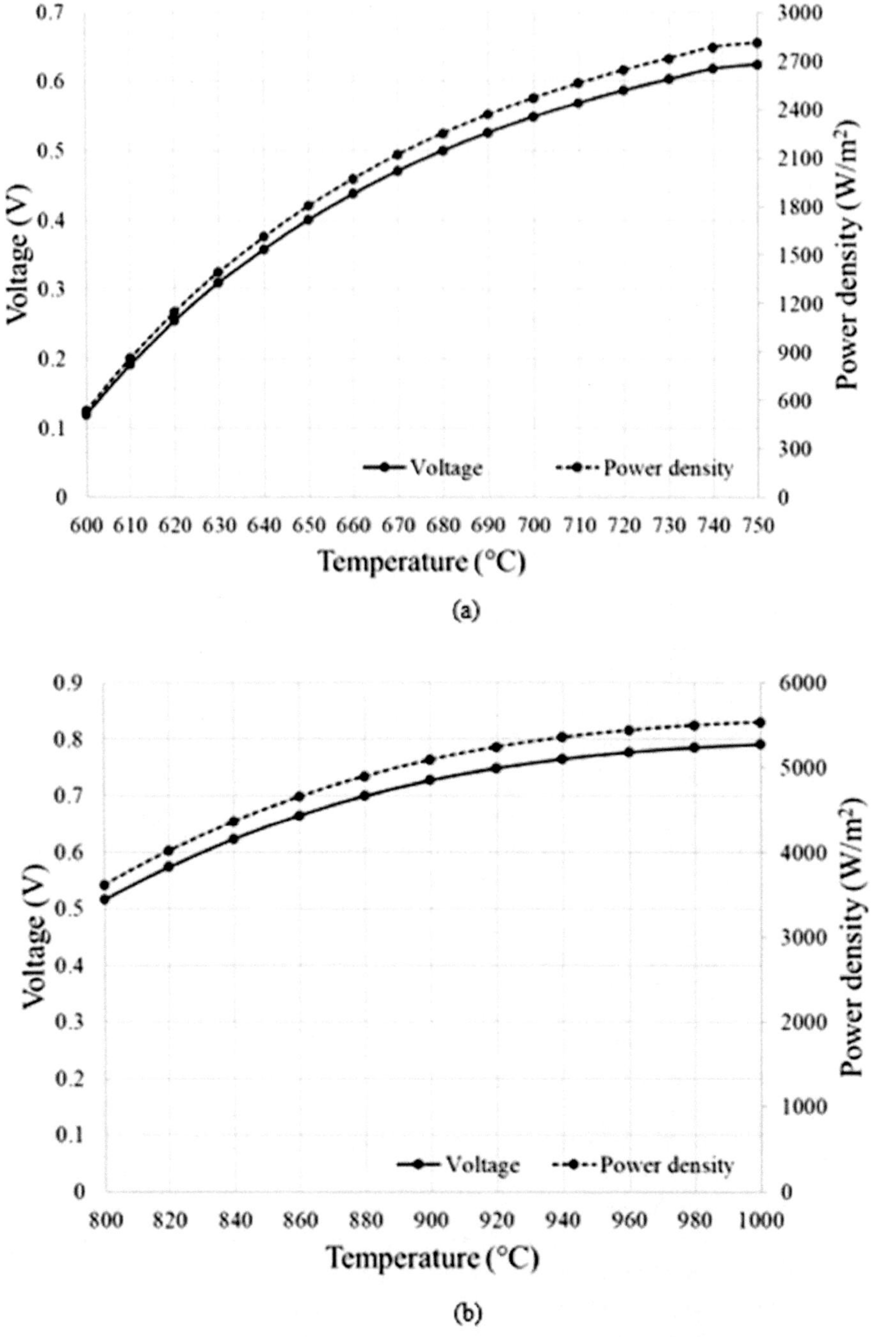

Fig. 4 Effect of temperature on voltage and power density of (**a**) MCFC; (**b**) SOFC

evaluation. We found that the integrated system of biogas tri-reforming with SOFC was more efficient than the integrated system with MCFC. However, MCFC operated at a lower temperature than SOFC. Lower temperature operation requires less energy usage, directly leading to financial productivity.

Acknowledgments The authors acknowledge the financial support from the Engineering Faculty, Thammasat School of Engineering, Thammasat University.

References

1. Department of Alternative Energy Development and Efficiency, Map of biogas status for production and usage in Thailand, http://webkc.dede.go.th/testmax/node/671, last accessed 2022/02/17.
2. Stephen, J., Aarva, A., Devianto, H., Bove, R., Moreno, A.: SOFC and MCFC: Commonalities and opportunities for integrated research. International Journal of Hydrogen Energy 36(16), 10337–10345 (2011).
3. Manenti, F., Pelosato, R., Vallevi, P., Leon-Garzon, A., Dotelli, G., Vita, A., Faro, M., Maggio, G., Pino, L., Aricò, A.: Biogas-fed solid oxide fuel cell (SOFC) coupled to tri-reforming process: Modelling and simulation. International Journal of Hydrogen Energy 40(42), 14640–14650 (2015).
4. Santoni, F., Della Pietra, M., Pumiglia, D., Boigues Muñoz, C., McPhail, S., Cigolotti, V., Nam, S., Kang, M., Yoon, S.: Accurate in-operando study of molten carbonate fuel cell degradation processes -part I: physiochemical processes individuation. Electrochimical Acta 291, 343–352 (2018).
5. Zhang, Y., Cruz, J., Zhang, S., Lou, H., Benson, T.: Process simulation and optimization of methanol production coupled to tri-reforming process. International Journal of Hydrogen Energy 38(31), 13617–13630 (2013).
6. Longmali, J., Aggarangsi, P.: In-situ methane enrichment of palm oil mill wastewater by effluent recirculation to stripping column. Naresuan University Engineering Journal 12, 63–72 (2017).
7. Duan, L., Xia, K., Feng, T., Jia, S., Bian, J.: Study on coal-fired power plant with CO_2 capture by integrating molten carbonate fuel cell system. Energy 177(2), 578–589 (2016).
8. Chatrattanawet, N., Saebea, D., Authayanun, S., Arpornwichanop, A., Patcharavorachot, Y.: Performance and environmental study of a biogas-fuelled solid oxide fuel cell with different reforming approaches. Energy 146, 131–140 (2018).
9. Baldinelli, A., Barelli, L., Bidini, G.: Upgrading versus reforming: an energy and exergy analysis of two solid oxide fuel cell-based systems for a convenient biogas-to-electricity conversion. Energy Conversion Management 138, 360–374 (2017).

Part II

Agricultural Ecological Environment and Safe Production

Identification of Community-Level Sustainability Indicators: A Climate Change and Risk Management Perspective

Hari Chandana Ekkirala, Debasish Brahma, A. R. Amritesh, Fernanda Imada de Lima, Maneesha V. Ramesh, Balmukund Singh, Maxwell Samaila, and Prince Destiney Ugo

Abstract

Measuring the sustainability of a community at a local level is fast gaining global prominence amongst policymakers, sustainability thinkers, and local stakeholders. It is crucial to measure sustainability to locate our present condition relative to conditions that have prevailed in the past and to develop relevant perspectives to frame actions for the future. Various indices have been used to measure sustainability. Every measure has its strengths and limitations, primarily due to sustainability's contextual nature. This paper looks at sustainability from a climate change and risk management perspective and studies vulnerability as a sustainability indicator for an agroforestry ecosystem for two communities, Chandrapuri, Uttarakhand, and Guptapada, Odisha.

Keywords

Agroforestry ecosystems · Social vulnerability · Climate change · Risk management

H. C. Ekkirala (✉) · D. Brahma · A. R. Amritesh · F. Imada de Lima · M. V. Ramesh · B. Singh · M. Samaila · P. D. Ugo
Amrita School for Sustainable Development, Amrita Vishwa Vidyapeetham, Kollam, Kerala, India
e-mail: amiddids20008@am.students.amrita.edu

J. T. Anderson (ed.), *Advances in Environmental Sustainability*, Springer Proceedings in Earth and Environmental Sciences,
https://doi.org/10.1007/978-3-031-26365-1_4

1 Introduction

Developmental initiatives concentrated on remote communities have faced numerous challenges. The primary concern grappling with today's policymakers is how to make this development sustainable. Sustainability concerns our ability to maintain a coupled human-nature system at a desirable state for multiple generations in the face of anthropogenic and environmental perturbations and uncertainties. It is essential to measure sustainability to locate our present condition relative to conditions that have prevailed in the past and to develop relevant perspectives to frame actions for the future if we are to realize this objective. At the very least, we need to adequately define what is to be sustained, what is to be developed, and how these components should be balanced or integrated. In the context of sustainable development, these performance metrics are primarily Sustainability Indicators and Indices (SIIs), which have played an increasingly important role in the science and practice of sustainability.

The proposed communities for a village-level comparison for this study were initially five. These villages were selected as they strongly rely on the agroforestry sector for livelihood and economy through fodder and farming. These villages also provide regulatory ecosystem services regarding carbon stock and soil formation. As part of the Live In Labs®, Amrita Vishwa Vidyapeetham has adopted numerous villages across India to work on various sustainable development goals. The villages selected for this study are identified through this program are plagued by Climate Change and Risk Management issues in the agroforestry sector. These villages are Chandrapuri and Udalka in Uttarakhand; Guptapada in Orrisa; Sarai Nooruddinpur in Uttar Pradesh; Anthonier Colony in Munnar, Kerala. However, the research presented here has been narrowed to Chandrapuri in Uttarakhand and Guptapada in Orissa. We chose these villages because although they are geographically and culturally diverse, they have similar ecosystem trade-offs and biophysical and social drivers.

2 Justification for the Use of Sustainable Development Indicators and Indices at a Community-Level

Anthropogenic activities impact the ecosystem and need to be integrated into initiatives for ecosystem service analysis and restoration [1]. Similarly, merging rural development and Environmental Protection policies is vital to harmonize bottom-up and top-down approaches [2]. It is important to empower communities at the bottom of the pyramid by implementing technological interventions in conjunction with community co-design frameworks[3]. In this regard, understanding vulnerability as a sustainable development index for rural agroforestry communities is essential now more than ever due to the joint impacts of COVID-19 and Climate Change on these ecosystem services [4, 5].

It is stated that, "an indicator refers to a variable or an aggregate of multiple related variables whose values can provide information about the conditions or trajectories of a system or phenomenon of interest" [6]. Based on the above definition of indicators and indices, crafting thus at the community level is paramount in ensuring sustainability. Community-level indicators and indices are the ultimate outcomes of an initiative. The community uses them to determine and measure the impact of changes happening in the course of implementation. Community-level indicators and indices provide a bottom-top approach to issues while giving a broader perspective. This approach brings inclusivity with a shared desire to focus on the ultimate goal. Community-level indicators and indices help the community to develop a collective action plan and push for its implementation. Consequently, it allows the community to measure progress towards a common goal.

3 Measuring Vulnerability for Agroforestry Ecosystems: A Climate Change and Risk Management Perspective

3.1 Identifying the Indicators that Are Specific to the Selected Context

The inventory of Sustainable Development tools has been cataloged into temporal characteristics, focal or coverage areas, and integration of the nature-society systems. It is further explained that indicators are simple measures with a broad scope that are sensitive to change and allow for quantifications [7]. An indicator is defined as "the operational representation of an attribute (quality, characteristic, and property) of a given system, by a quantitative or qualitative variable (for example, numbers, graphics, colors, and symbols), including its value, related to a reference value" [8]. When a set of indicators are quantified and aggregated, they become a measurement index.

For practical purposes, the system's vulnerability is studied as an indicator of sustainability and compared to theme-based approaches for measuring sustainability. Thus we are looking at Climate Change and Risk Management perspectives specifically for community-level agroforestry services. The paper recognized that there could be various other means of measuring sustainability, but vulnerability has been chosen as a practical indicator from a Climate Change and Risk Management perspective. The framework detailed here is adapted from [9]. These indicators also function as metrics for social, environmental, economic, and institutional development within communities.

The assessment method applied was vulnerability analysis for the agroforestry sector for the two study regions. This intends to evaluate the status, drivers, and possible trade-offs between socio-ecological systems, which in this case are the natural capital and farm capital. Vulnerability is "an aspect that is driven by inadvertent or deliberate human action that reinforces self-interest and the distribution of power in addition to interacting with physical and ecological systems" [10]. Vulnerability has three dimensions: exposure, sensitivity, and adaptive capacity.

Table 1 Definitions and use of components under vulnerability analysis

Dimension	Component	Definition/usage
Adaptive capacity	Decision support	Socio-economic agency in pre, post, and during a disaster event
	Social support	Access to assistance from within the community and external agencies
	Economic support	Level of assets and economic potential
	Livelihood strategy	Income level and job security
	Farm support	Access to external information and guidance on adapting farm practices to climate change
	Nutrition and health	Impacts of climate change and extreme weather events on nutrition and health
Exposure	Climate and extreme weather events	Inventory of temperature, rainfall, and natural capital
	Impact of extreme weather events	Damage to agriculture infrastructure, livelihood, and health
Sensitivity	Human wellbeing	Economic challenges in the aftermath of a hazard or due to climate change
	Natural capital	Loss of ecosystem as a source of livelihood

Adapted from Gupta et al. (2019)

- Exposure: degree of climate stress on a unit of analysis (long-term means, variability, extreme events).
- Sensitivity: the degree to which a system will respond to a change in climate (positively or negatively); for example, Are you a farmer? Do you depend on corals? Do you like the heat?
- Adaptive capacity: the ability of a system to adjust and adapt to actual or expected climate stresses (related to wealth, technology, skills, access to resources, and management capabilities).

The resulting index will help determine the system's sensitivity and adaptivity to Climate Change and extreme weather events impacts. This assessment's temporal aspect is prospective, aiming to forecast the future effects. The focal area is an integrated assessment geared towards product and policy monitoring regarding Climate Change adaptation and disaster risk reduction. Under each of the three dimensions of exposure, sensitivity, and adaptive capacity, we aggregated additional components (Table 1).

3.2 Identification of the Measurement Methods

The approach used to define a measurement metric for each component discussed above is to identify the question that best captures the indicator. Thresholds for all indicators have not been determined as that would require pilot data collection to understand the relevance of each indicator from a community member's perspective. Further analysis and understanding of socio-ecological interrelationships for each village are needed to identify the thresholds that cater to particular communities.

A one size fits all approach is insufficient in measuring Climate Change impacts and extreme weather events on the Agroforestry Ecosystem services. The practical application of some of these indicators might vary as each village has differing aspects, such as significant sources of livelihood, economic sectors present, ecosystems present, drivers, and trade-offs (Table 2).

4 Derivation of Indices

The approach to aggregating the indicators and deriving an index is done through three levels. The first level, Level A pertains to indices related to decision support, social support, economic support, social network, livelihood strategy, farm support, food and health, climate and extreme weather events, impact of extreme weather events, and human well-being. These, when consolidated, give us information about the three dimensions of vulnerability which is the second level (adaptive capacity, exposure, and sensitivity). This information when consolidated leads to the final index for vulnerability analysis.

This is achieved by assigning a number from 1 to 5 to each question based on the interviews with village coordinators for both locations. The values were calculated against the threshold on the Likert scale. For each of the indicators, a certain value is indicated as the threshold based on the secondary research. The higher the value assigned the less impact it has on increasing vulnerability. Hence, a high value is preferred. For example, if they have steady electricity throughout the day, the value is 5. If they are highly exposed to extreme weather events, the value assigned is 1.

Each question is assigned a priority between 1 and 5 based on the requirements to tackle Climate Change adaptation and disaster risk reduction. For example, the availability of electricity was given a priority of 5, and the use of renewable energy sources was given a priority of 3. The formula of weight = value*priority was used to arrive at the weight value for each indicator. The cumulative average weight value for each component within the dimensions of exposure, adaptive capacity, and sensitivity was calculated (such as decision support, farm support, human wellbeing, etc.). Additionally, a visualizer was created with both the villages to compare the two. If a component has a high value, that would mean that it has less effect on increasing vulnerability and if an element has low value, those are areas of concern within the village, and interventions need to be made to address indicators within that component to reduce vulnerability.

Table 2 Measurement metrics for each indicator

Dimension	Component	Indicator	Metrics of measurement
Adaptive Capacity	Decision support	Physical agency	Age, physical and mental handicap, temporary illnesses, pregnant
		Access to indigenous knowledge	Membership and participation in social and community level organizations/decision-making power
		Gender representation	50% female representation from the community or one female represented per household
		Education level	Minimum high school diploma
	Social support	House type	Concrete or thatched houses
		Electricity availability	Hours of supply
		Sanitation facility	Toilet in the household/ common toilets in village/ open defecation
	Economic support	Residential land ownership	Amount of land owned
		Farm land ownership	Amount of land owned
		Livestock	The number of cattle and poultry owned.
		Availability of non-renewable sources of energy	Sources used
		Availability of renewable sources of energy	Sources used
		Drinking water availability	Sufficient access to drinking water
		Drinking water sources and accessibility	Access to borewells and check dam/distance to the source
	Social network	External assistance	Access to NGO support, social volunteers, and vocational training
		Community assistance	Awareness programs, sarpanch meetings, self-help groups
	Livelihood strategy	Job security	Permanent jobs, ad-hoc employment, urban repatriation
		Income security	The number of months employed
		New activity	Adult literacy programs

(continued)

Table 2 (continued)

Dimension	Component	Indicator	Metrics of measurement
	Farm support	Crop advisory and management	Access to knowledge on meteorological dept. weather information, early warning, cultivating multiple types of crops to adapt to different weather conditions, and access to expert opinion. Access to weather information, advisory on irrigation schedules, crop disease monitoring, insurance awareness, etc.
		Farm nutrient usage	Number of farmers using chemical fertilizers/ acres of land using chemical fertilizers/number of farmers using chemical pesticides
		Irrigation support	Shift to irrigated crops, dams, cannel, water channels front the river/adjustment for irrigation water/adjustment for water scarcity
	Food and health	Nutrition and food impacts on food availability	Per capita food consumption and body mass index/percentage of food production in the village/months during which cultivated food lasts
		Access to healthcare	Access to government hospitals, vaccines, mobile hospitals, ambulances, etc.
Exposure	Climate and extreme weather events	Temperature	Days of maximum and minimum temp.
		Rainfall	Number of monsoon months (May to Sept, June to October)
		Drought	Amount of rainfall
		Cloudbursts	Number of early warnings issued/number of incidences reported
		Landslides	Number of early warnings issued/number of incidences reported
		Forest fires	Number of early warnings issued/number of incidences reported
		Glacial Lake Outbursts	Number of early warnings issued/number of incidences reported
		Cyclones	Number of early warnings issued/number of incidences reported
		Heatwaves	Time period
		Cold spells	Time period
		Agriculture productivity	Amount of monetary loss in damage due to land loss
	Impact of extreme weather events	Water availability	Number of damaged water sources
		Health	Number of injured people/lives lost
		Land fragmentation	Area of deforestation/change in land use patterns

(continued)

Table 2 (continued)

Dimension	Component	Indicator	Metrics of measurement
Sensitivity	Human wellbeing	Climate change	Number of extreme events per year experienced by the village
		Tourism	Number of tourists per season
		Road construction	Amount of new road widenings/road access to the village
		Hydro construction	Number of new dams/access to drinking water
		Wildlife conflict	Number of households affected in a month
		Fuelwood collection	Number of trees removed
	Natural capital	Fodder collection (forest)	Hectares of forest land/floor removal
		Soil fertility	Area of fertile soil sufficient for crop production

Adapted from Gupta et al. (2019)

5 Discussions and Conclusions

5.1 Visualizer for the Index and Inferences

The results obtained for each component are discussed in this section based on the adopted methodology. Figure 1 illustrates the community-wise comparison of the vulnerability index in Chandrapuri, Uttarakhand and Guptapada, Odisha. The decision support component has a value of about 9.75 and 8.75 in Chandrapuri and Guptapada, respectively. The value for physical agency linked to hospital care scored the lowest in both locations due to inadequate healthcare facilities. Gender representation and female financial independence were also low on the list. Social support has very similar weighted values in both villages with about 16.67 and 16.33, respectively. Furthermore, in Guptapada, most houses are made of thatched roofs while in Chandrapuri they are mostly concrete. Electricity supply is fluctuating in Chandrapuri, while it is steady in Guptapada. Both villages have sanitation facilities with a toilet in each household. Economic support scored a value of 9 in Chandrapuri and 12.5 in Guptapada. They are highly dependent on non-renewable sources of energy like petroleum and firewood. They own cattle in both villages and everyone owns land in Chandrapuri. However, one of the key concerns identified is that they lack a social support network in Chandrapuri, with a weighted value of 9 as compared to Guptapada, with a weighted value of 12.5. Both locations have a value of 6.33 for livelihood strategy. Both locations lack access to vocational training or adult literacy programs.

Both locations lack external support in terms of crop advisory. However, they rely on ancestral practices and do not invest in pesticides or fertilizers. They use manure from household cattle. Both villages scored a value of 6 for farm support.

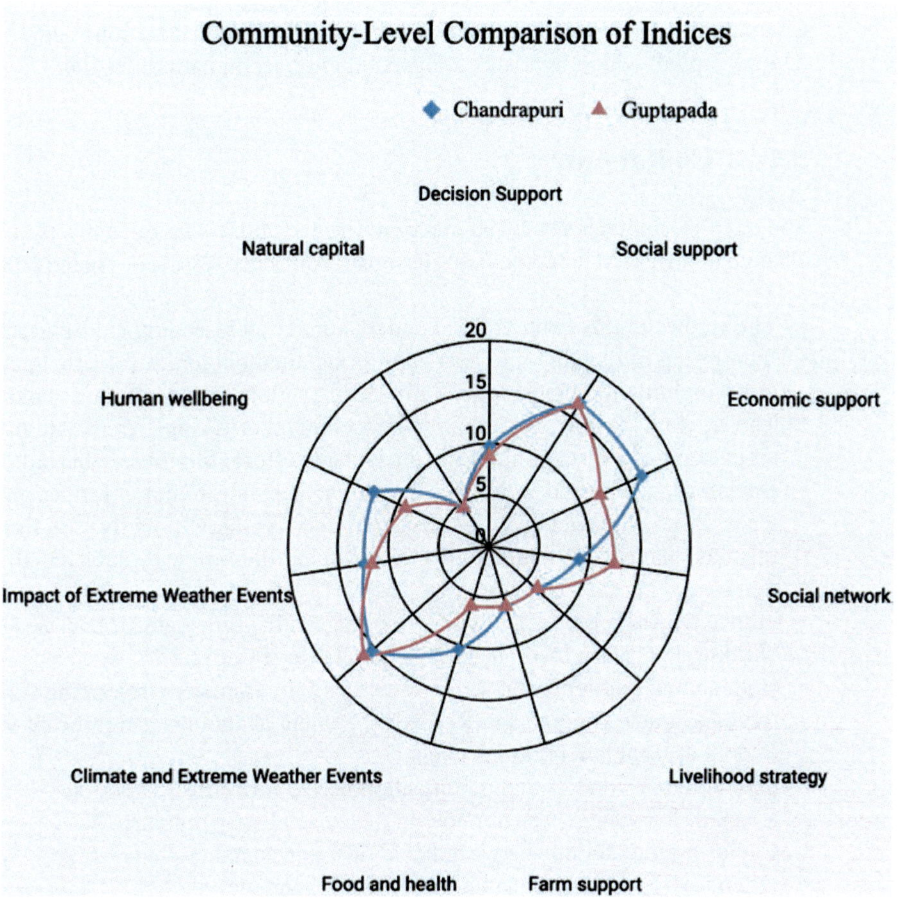

Fig. 1 Community-wise comparison of vulnerability index in Chandrapuri, Uttarakhand and Guptapada, Odisha

The population is healthy and dependent on paddy, gehu, and Dhan in Chandrapuri. Both locations have less than adequate healthcare facilities. Futhermore, Guptapada scored a weight value of 6 while Chandrapuri scored 10.50. Chandrapuri got a value of 15.45 for Climate Change and weather extremes. This region is highly prone to landslides, earthquakes, cold spells, cloudbursts, and floods. Guptapada scored a slightly higher value of 16.36. Additionally, this region is prone to forest fires, extreme rainfall events, and cyclones. Both villages incur heavy losses in terms of agricultural yield due to natural hazards.

Chandrapuri scored a slightly higher value for impacts of extreme weather events with a score of 12.5, as they face recurring short supplies of water and electricity due to landslides. The effects of Climate Change impact land fragmentation more in Guptapada, which scored 11.67. Furthermore, Chandrapuri scored 12.67 for human wellbeing, and Guptapada scored 9.17. Also, Chandrapuri relies heavily on tourism, whereas Guptapada does not see many tourists. In addition, Guptapada also lacks proper transport corridors. It relies heavily on fuelwood collection and, similar to Chandrapuri, witnesses extensive wildlife conflict. Both locations scored 4.5 for natural capital, which indicates a substantial impact on natural capital.

5.2 Limitations

Some of the challenges incurred in measuring and addressing the impacts of Climate Change and Risk Management on the agroforestry ecosystem are listed below:

- The studies in this context often underestimate socio-ecological Vulnerability.
- Identifying or drafting the most appropriate methodologies for data aggregation and weighting indicators, especially during a global pandemic, was challenging.
- The existing frameworks are more concentrated on a single event and often lack an in-depth understanding of the interrelated effects of multiple hazards.
- Decision-makers and policymakers often lack on-ground information, which leads to mismanagement of ecosystem services. Additionally, the impacts of mismanagement are multi-fold in developing nations with substantial population loads.
- Often, they are more involved in coping with immediate impact and neglect building long-term resilience to Climate Change impacts.
- Both spatial and temporal scales are essential parameters in looking for indicators of Climate Change and Risk Management as there is a significant variation in both at local and regional levels.
- Often, there is a lack of communication between various stakeholders because of a lack of knowledge transfer among interdisciplinary domains.
- Capturing interactions between ecological and social systems.
- Often, there is a lack of community-centric approaches.
- This leads to a need for a more flexible and dynamic approach.

6 Conclusion

Climate Change can impact all walks of life but measuring its magnitude at various levels is complex. However, this measurement is essential for quicker recovery, disaster resilience, and long-term sustainability. To calculate the Climate Change impacts and preparedness of the community, we have identified key indicators and prepared in-depth metrics of measurement, which are utilized to collect data from two villages in India. This will further be improved and incorporated into a comprehensive framework for ecosystem service monitoring. Based on the framework detailed in this paper, the data collected from the other three villages of Udalka, Uttarakhand; Sarai Nooruddinpur, Uttar Pradesh; and Anthonier Colony, Munnar will also be analyzed and the results will be published in future work.

Acknowledgments The E4LIFE International Ph.D funded this project. We thank the Fellowship Program offered by Amrita Vishwa Vidyapeetham. We extend our gratitude to the Amrita Live-in-Labs® academic program for providing all the support. We also thank our chancellor, Sri Mata Amritanandamayai Devi, for her exemplary vision focusing on selfless service and compassion to tackle complex sustainable development problems in communities across India.

References

1. Lengefeld, E., Metternicht, G. and Nedungadi, P., 2020. Behavior change and sustainability of ecological restoration projects. *Restoration Ecology*, *28*(4), pp.724–729.
2. Lengefeld, E., Stringer, L.C. and Nedungadi, P., 2021. Livelihood security policy can support ecosystem restoration. *Restoration Ecology*, p.e13621.
3. M. V. Ramesh, R. Mohan and S. Menon, "Live-in-Labs: Rapid translational research and implementation-based program for rural development in India," *2016 IEEE Global Humanitarian Technology Conference (GHTC)*, 2016, pp.164–171, https://doi.org/10.1109/GHTC.2016.7857275.
4. Habanyati, E.J., Paramasivam, S., Seethapathy, P., Jayaraman, A., Kedanhoth, R., Viswanathan, P.K. and Manalil, S., 2022. Impact of COVID-19 on the Agriculture Sector: Survey Analysis of Farmer Responses from Kerala and Tamil Nadu States in India. Agronomy, *12*(2), p.503.
5. Angom, J., The Potential role of Green Finance in realizing Sustainable recovery from COVID-19 and achievement of Sustainable Development Goals. International Journal of Research and Innovation in Social Science (IJRISS) |Volume V, Issue IV, April 2021|ISSN 2454–6186
6. Wu, J. and Wu, T., 2012. Sustainability indicators and indices: an overview. Handbook of sustainability management, pp.65–86.
7. Ness, B., Urbel-Piirsalu, E., Anderberg, S. and Olsson, L., 2007. Cat- categorizing tools for sustainability assessment. Ecological Economics, 60(3), pp.498–508
8. Waas, T., Huge, J., Block, T., Wright, T., Benitez-Capistros, F. and Ver- bruggen, A., 2014. Sustainability assessment and indicators: Tools in a decision-making strategy for sustainable development. Sustainability, 6(9), pp.5512–5534.
9. Gupta, A.K., Negi, M., Nandy, S., Alatalo, J.M., Singh, V. and Pandey, R., 2019. Assessing the vulnerability of socio-environmental systems to climate change along an altitude gradient in the Indian Himalayas. Ecological Indicators, 106, p.105512
10. Adger, W.N., 2006. Vulnerability. Global environmental change, 16(3), pp.268–281

Study of Extreme Precipitations in Corn Culture in Chupaca, Junin – Peru 2009–2019

Betsy L. Argomedo, Ingrid S. Galván, Diego A. Gaspar, Iralmy Y. Platero, and Michelle A. Mollehuara

Abstract

Using 11 years of data in Chupaca, Junin region, Peru, we studied the effects of extreme precipitation on the maize crop. For the project, extreme annual precipitation and temperature data were obtained from the Viques and Huayao meteorological stations of the National Meteorological and Hydrological Service of Peru, as well as crop yield and production data. The Spearman correlation method was used to carry out the research. The results indicated that in 2014 there was a higher yield of corn, and in 2011, there was the highest amount of rainfall. We conclude that extreme rainfall does not influence the yield of corn in the period under study.

Keywords

Crop yield · Precipitation · Temperature

1 Introduction

The Mantaro Valley is highly vulnerable to extreme weather events, according to the Servicio Nacional de Meteorología e Hidrología (SENAMHI) of Peru, such as extreme precipitation, which harms food security and the development of the economy of society. During the 1995–1996 and 2010–2011 agricultural seasons, frosts, veranillos, and heavy rains caused floods, landslides, and fungal diseases that

B. L. Argomedo (✉) · I. S. Galván · D. A. Gaspar · I. Y. Platero · M. A. Mollehuara
Professional Academic School of Environmental Engineering, Universidad Continental, Huancayo, Peru
e-mail: 73500700@continental.edu.pe

J. T. Anderson (ed.), *Advances in Environmental Sustainability*, Springer Proceedings in Earth and Environmental Sciences,
https://doi.org/10.1007/978-3-031-26365-1_5

damaged agriculture [1]. In 2016, the government declared Mantaro Valley an agricultural emergency because 6400 hectares of crops were affected by heavy rains, hailstorms, and drought [2]. That same year the regional director of Agriculture - Federico Ollera, pointed out that these meteorological events had generated the loss of around 55 million Peuvian soles [3].

In the province of Chupaca, planting begins in September and harvests in March since the rainy season is generally between November and March, and the dry season between April and October [4].

Corn is an important staple food for humans and essential food for animals. Yields may change due to increased precipitation extremes [5].

Our research explores the relationship between environmental temperature, precipitation extremes, and corn yields, it is suggested that these variables may be directly related to corn yields. Corn is a primary raw material for manufacture of thousands of industrial products worldwide [6]. The production of corn represents an essential economic contribution in several countries in the Americas, so climate change directly impacts on the efficiency of its production and the crop's adaptability in terms of its distribution [5].

Studies have shown the negative impact of higher temperatures on corn production [7]. This is partly because the effects of climate change on maize yields differ spatially.

2 Methodology

Currently, in the province of Chupaca, there is a population of 52,988, of which 20,633 people live in rural areas according to the census conducted in 2017 by Instituto Nacional de Estadística e Informática (INEI) [8].

The effects of crops were studied using data from 2009 to 2019, such as temperature and extreme precipitation from the meteorological stations of Viques and Huayao of SENAMHI [9].

Our study occurred in three representative districts in the Chupaca Province (Fig. 1).

Chupaca had the highest maize cultivation in the three (Table 1).

According to the Dirección Regional de Agricultura de Junín (DRAJ), corn production in the three study districts from 2009 to 2019 generally declined (Table 2).

In the research for the collection techniques, the data obtained by SENAMHI from the automatic weather stations were used, in addition to securing yield data from the Dirección Regional de Junín (DRAJ). We analyzed maize crop yield data using Microsoft Excel, ArcGIS and SPSS.

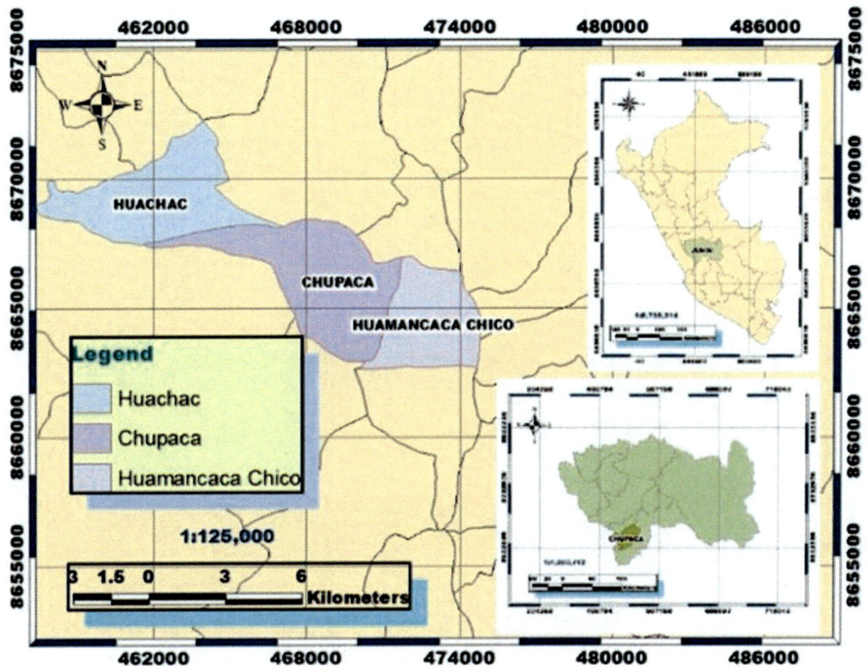

Fig. 1 Delimitation of the study area

Table 1 Hectares of maize cultivation for the sample

District	Hectares
Chupaca	120
Huachac	70
Huamancaca Chico	32

3 Results

3.1 Corn Crop Yields

The current crop yield represents variations due to climatic changes according to data taken from SENAMHI [9]. From 2009 to 2016, there was increased corn production, and, in recent years, there was less production due to weather events (Fig. 2).

There were lower corn yields in the last two years (Fig. 2). Corn is a critical crop in the province of Chupaca. In 2014, there were higher yields in both Chupaca and Huachac and Huamancaca Chico with 12,583, 12,321, and 12,393 kg/ha, respectively, having an overall yield of 12,470 kg/ha, and a lower yield in 2019 with 10,531 kg/ha.

Table 2 Maize crop production

Year	Chupaca	Huachac	Huamancaca Chico
2009	3.487	1.593	1.209
2010	3.360	1.905	1.326
2011	3.121	1.584	1.268
2012	2.885	1.426	1.183
2013	2.847	1.526	1.193
2014	2.743	1.663	719
2015	2.781	1.659	643
2016	2.587	1.526	578
2017	1.650	761	136
2018	1.951	1.161	461
2019	1.291	714	334

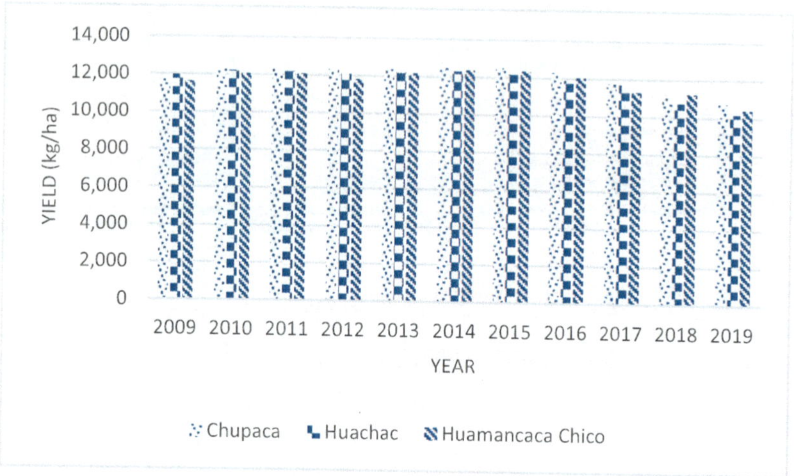

Fig. 2 Annual corn crop yields according to districts in the province of Chupaca

3.2 Precipitation and Temperature

The extreme precipitation data used were from the Viques and Huayao meteorological stations. The highest amount of precipitation (mm) occurred in 2011, the lowest amount of rainfall in 2016, and the lowest and highest temperatures (°C) appeared in 2018 and 2016, respectively (Figs. 3 and 4).

Figure 3 shows the extreme precipitation data used for the Viques meteorological stations from 2009 to 2019. The highest amount of precipitation was recorded in 2011 with 1124.2 mm and the lowest in 2016 with 559.8 mm. The lowest temperature was recorded in 2018 at 12 °C and the highest in 2016 at 13.5 °C.

The extreme precipitation data from the Huayao meteorological station varied from 2009 to 2019 (Fig. 4). The highest amount of precipitation was recorded in 2011 with 912.1 mm and the lowest in 2016 with 559.8 mm. The lowest temperature was recorded in 2018 at 11.1 °C, and the highest in 2016 at 12.3 °C.

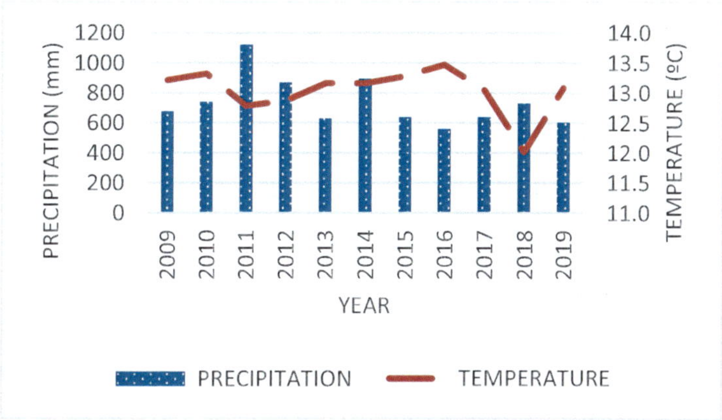

Fig. 3 Climogram of Viques

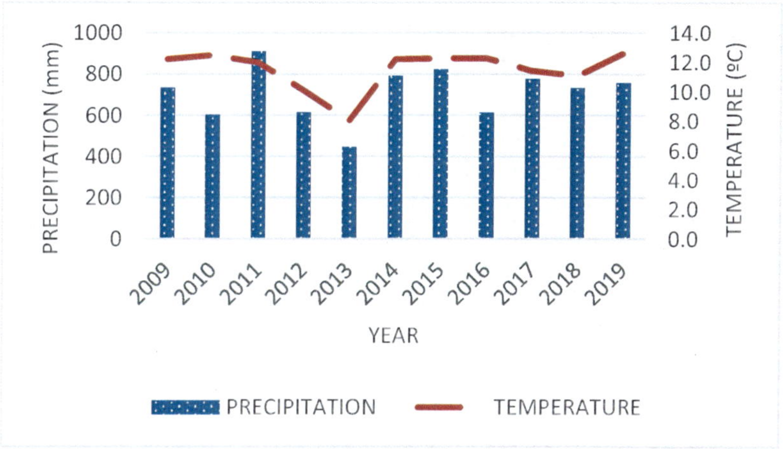

Fig. 4 Climogram of Huayao

3.3 Average Annual Precipitation

The following is the average annual rainfall for each month for 11 years, comparing the two stations of Viques and Huayao.

Figure 5 shows the precipitation data at the Viques weather station from 2009 to 2019, in the "X" axis marks the precipitation in mm, and the "Y" axis shows that the average months where there was more precipitation was from December to March as well as the middle months had low levels of precipitation coinciding with the data from the Huayao station. However, in this station, there was not much dispersion in the data, especially in December.

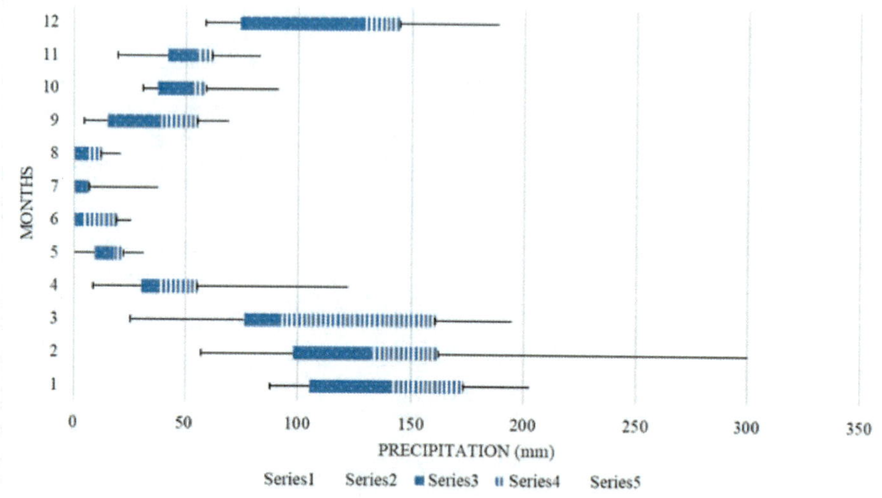

Fig. 5 Box and whisker plot of the Viques weather station

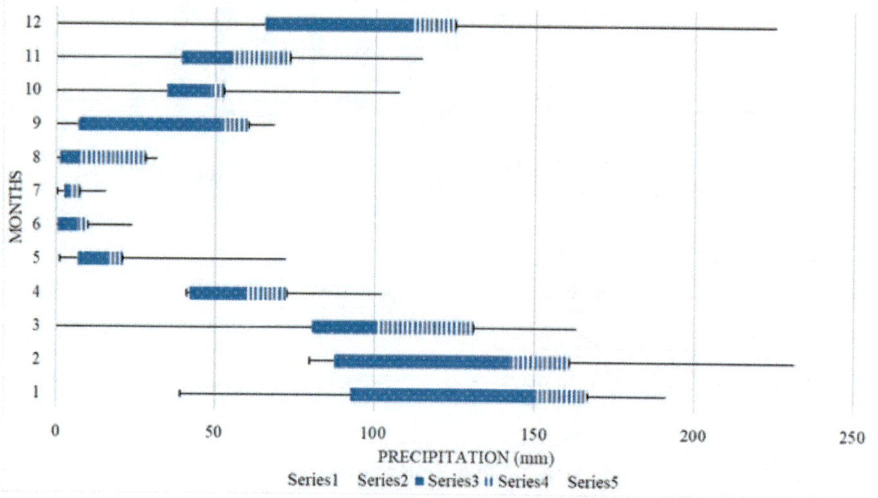

Fig. 6 Box and whisker plot of the Huayao weather station

Figure 6 represents the precipitation data at the Huayao weather station between the years 2009 and 2019, in the "X" axis marks the precipitation in mm, and in the "Y" axis the months of the year, showing that the rainy period in the study area is from December to March. June and July are less rainy, and the data are very scattered in the month of December because, in the year 2013, no values were recorded.

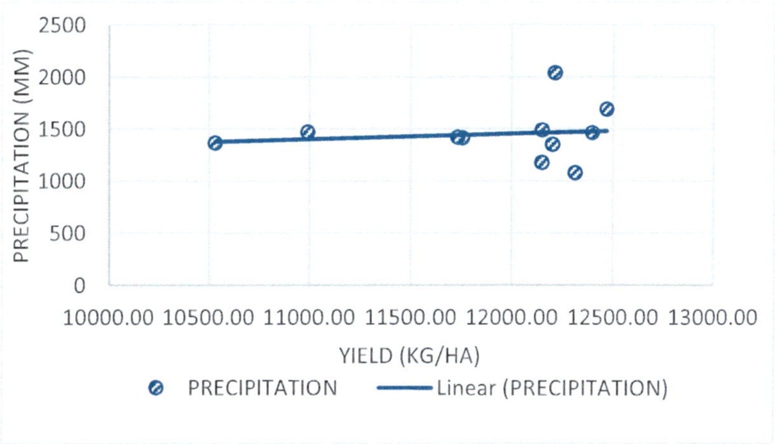

Fig. 7 Scatterplot of precipitation and corn yields

3.4 Spearman Method

Extreme rainfall varies between 20 and 25 mm/h in the form of showers and is often accompanied by hail [10]. Rain directly affects the yield of the corn crop since rainfall is optimal between 500 and 700 mm in its phenological stages: emergence, leaf development, ear, milky ripening, and pasty ripening [11] (Fig. 7).

The Spearman method in SPSS was used, which determined no correlation between rainfall and corn yield in the detailed study area (P = 0.519).

Figure 7 shows the dispersion diagram of rainfall (mm) and corn yield (kg/ha) worked in excel. An increasing trend line can be seen, representing the linear correlation between both variables. It can be seen that the points are scattered because some years have low rainfall, thus resulting in no correlation between the yield and rainfall variables. The Spearman test between temperature and yield, at 95% confidence, indicated no correlation between temperature and yield of the corn crop (P = 1.0).

The diagram of dispersion of temperatures (°C) and yield of the corn crop (kg/ha) worked in excel can be appreciated, where a line of decreasing tendency can be visualized that represents the linear correlation between both variables, and it is observed that the points are dispersed because some years present low temperatures. Therefore it gives a result that there is no correlation between the variables of yield and temperature (Fig. 8).

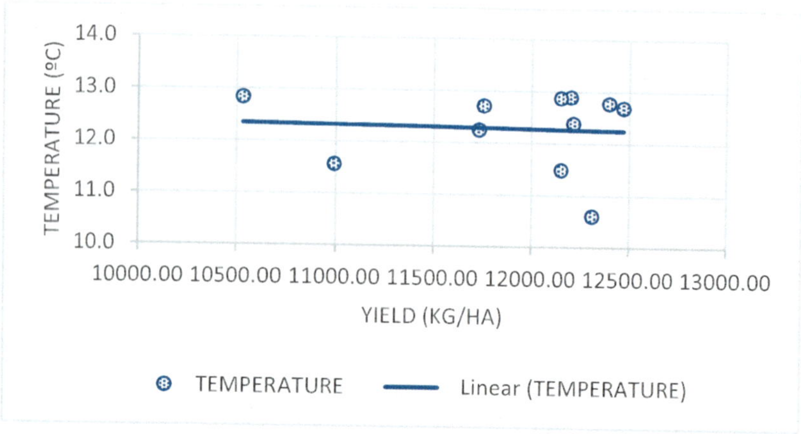

Fig. 8 Scatterplot of temperature and corn yields

4 Conclusions

The research work demonstrates the interest in climate change and its effect on maize crop in the selected districts of the province of Chupaca, concluding that the data obtained using the Spearman method shows that there is no relationship between rainfall and corn crop yield between 2009 and 2019. Furthermore, there is no relationship between temperature and corn production.

References

1. Instituto Geofísico del Peru. Manejo de riesgos de destrastes ante eventos metereologicos extremos en el valle del Mantaro. Primera ed. Martinez AG, editor. Lima, Peru: Lettera Gráfica SAC; 2012.
2. El comercio. Declaran en emergencia la agricultura en 9 provincias de Junin – Peru. El comercio. 2016 Febrero.
3. Tinoco P. Heladas y sequias en el valle del Mantaro causan perdidas por S/55 millones. Diario Andino. 2016 Noviembre.
4. Weather Spark. Weather Spark. [Online].; 2021. Available from: https://es.weatherspark.com/y/22266/Clima-promedio-en-Chupaca-Per%C3%BA-durante-todo-el-a%C3%B1o#:~:text=En%20Chupaca%2C%20los%20veranos%20son,m%C3%A1s%20de%2023%20%C2%B0C.
5. Mallma Carrera GA. Evaluación de la influencia del cambio climatico en los cultivos de Solanum Tuberosum y Zea Mays y en la microcuenca de Calicanto en Apurímac. Tesis para pregrado. Lima: Universidad Nacional Agraria La Molina, Lima; 2019.
6. WU Jz, ZHANG J, GE Zm, XING Lw, HAN Sq, SHEN C, et al. Impact of climate change on maize yield in China from 1979 to 2016. Journal of Integrative Agriculture. 2021 Enero; 20(1).
7. Arias Hernandez G. Efecto del cambio climatico en el rendimiento de maiz (Zea mays) variedad jaguan para cuatro regiones del estado de Coahuila. Tesis. Mexico: Universidad Autonoma Antonio Narro, Saltillo; 2021.

8. Instituto Nacional de Estadística e Informática. Resultados Definitivos de los Censo Nacionales 2017. [Online].; 2018. Available from: https://www.inei.gob.pe/.
9. Servicio Nacional de Meteorología e Hidrología del Peru. Datos hidrometeorológicos. [Online].; 2021. Available from: https://www.gob.pe/senamhi.
10. Arroyo Aliaga J, Machuca Manrique E, Gurmendi Párraga P. Caracterización espacial de la frecuencia e intensidad de tormentas desde el satélite GOES-12 y la Estación Meteorológica del Observatorio de Huancayo. Apuntes de Ciencia & Sociedad. 2014 ENERO; 4(1).
11. Ministerio de Agricultura y Riego. Plan nacional de cultivos campaña agrícola 2019–2020 Acosta Reátegui JM, editor. Lima, Peru: Diseño & Edición Digital; 2020.

Spatiotemporal Mapping of Agroforestry Degradation of Five Districts of Uttarakhand, India with Spectral Indices

Susan V. Panicker, Ekkirala Hari Chandana, and Maneesha V. Ramesh

Abstract

Land degradation adversely affects socio-economic status and ecosystem services. Land sharing, land separation, and land conservation are part of environmental conservation and are gaining momentum to serve the needs of humans and climate regulation. Simultaneous environmental changes challenge biodiversity persistence and human well-being. Ecosystem degradation resulted in numerous social and environmental issues, such as extreme weather events, reduced production of forestry and agricultural-based livelihoods, and endangered biodiversity. This study aims to determine the spatial extent of agroforestry ecosystem degradation for two time periods, 2013 and 2021, in five districts of Uttarakhand, India, the parameters that led to the changes, and the importance of each of the parameters. By considering the heterogeneous nature of agroforestry degradation, identifying relevant parameters and spectral indices for classifications is a critical process in agroforestry ecosystem degradation analysis. Remotely sensed satellite images are used to evaluate the degradation and preparation of spectral indices. Understanding the changes and the underlying parameters help to deploy corrective measures to prevent further degradation and restore the ecosystem to provide functional outputs.

Keywords

GIS · Remote sensing · Spectral indices

S. V. Panicker (✉) · E. H. Chandana · M. V. Ramesh
Center for Wireless Networks & Applications (WNA), Amrita Vishwa Vidyapeetham
Amritapuri, Vallikavu, Kerala, India

J. T. Anderson (ed.), *Advances in Environmental Sustainability*, Springer Proceedings in Earth and Environmental Sciences,
https://doi.org/10.1007/978-3-031-26365-1_6

1 Introduction

Identifying and quantifying heterogeneous and complex parameters from remote areas are expensive and time-consuming. The data from the users are often incomplete and have limitations of spatial changes [1]. In India, overall precipitation decreased by about 68% [2], and there were multiple disasters in Uttarakhand, such as earthquakes, landslides [3], forest fires, and flash floods [4], and climate change increases these vulnerabilities. Five northwestern states of Uttarakhand, including the capital, were selected for the study based on the state's diverse geography, varying altitudes, and prevailing practice of agroforestry [5]. The highest net sown area is in Pauri (18.88%), followed by Tehri (14.81%), and Rudraprayag (7.59%) districts [6]. Uttarkashi is one of the famous tourist places, and Dehradun is the state's capital. The districts selected for the study are - Uttarkashi, Dehradun, Pauri Garhwal, Tehri Garhwal, and Rudraprayag. There was a flash flood in June 2013, one of the major disasters in Uttarakhand, which is one reason for selecting this period for the study.

India was the first country in the world to adopt an agroforestry policy in 2014. The National Agroforestry Policy of India is a comprehensive framework designed to improve agricultural livelihoods by maximizing agricultural productivity to mitigate climate change. It has pointed out how to promote agroforestry among various stakeholders [7]. Multiple Earth Observation (EO) techniques are available to evaluate the parameters, and selecting the appropriate methodology is essential in agroforestry degradation analysis. Disasters and anthropogenic activities are also important factors that lead to agroforestry land degradation [8], and these are to be considered while assessing ecosystem degradation. End-to-end analysis of interrelated physical and anthropogenic heterogeneous parameters influencing agroforestry degradation has not been conducted in one study, and reliable data sets and appropriate tools to accurately map the extent of degradation are missing. There is a lack of comprehensive research to identify the parameters, its interrelationships, spatiotemporal mapping of agroforestry degradation, and its impact on the provisioning of valuable ecosystem goods and services, considering the regional and local climatic scenarios. Understanding land use and land cover (LULC) [9] dynamics is necessary to understand the ecosystem changes and level of degradation. Agroforestry depends on rainfall and is an essential factor in high-altitude zones [10].

2 Methodology

Geographical information systems (GIS) and remote sensing (RS) techniques help to monitor the degree of degradation by evaluating the changes in LULC [11]. A multi-criteria analysis of relevant parameters and weighted matrix formulation was done to identify and rank the parameters. Weighted overlay analysis helped to analyze the degraded areas. For degradation analysis, six parameters – Slope, Elevation, LULC, NDVI, NDBI, and NDMI were considered [12] as shown in Fig. 1.

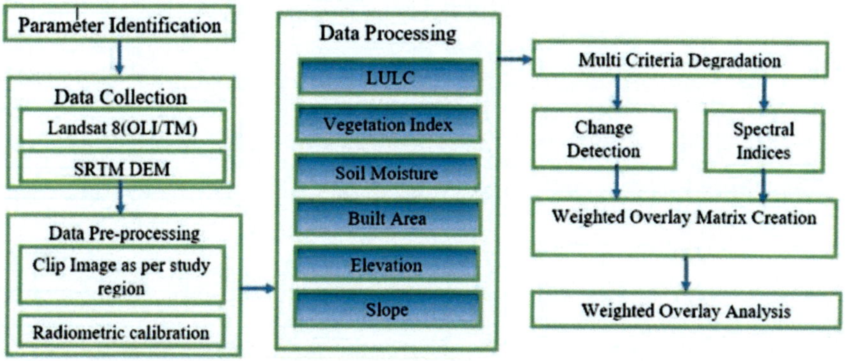

Fig. 1 Methodology block diagram

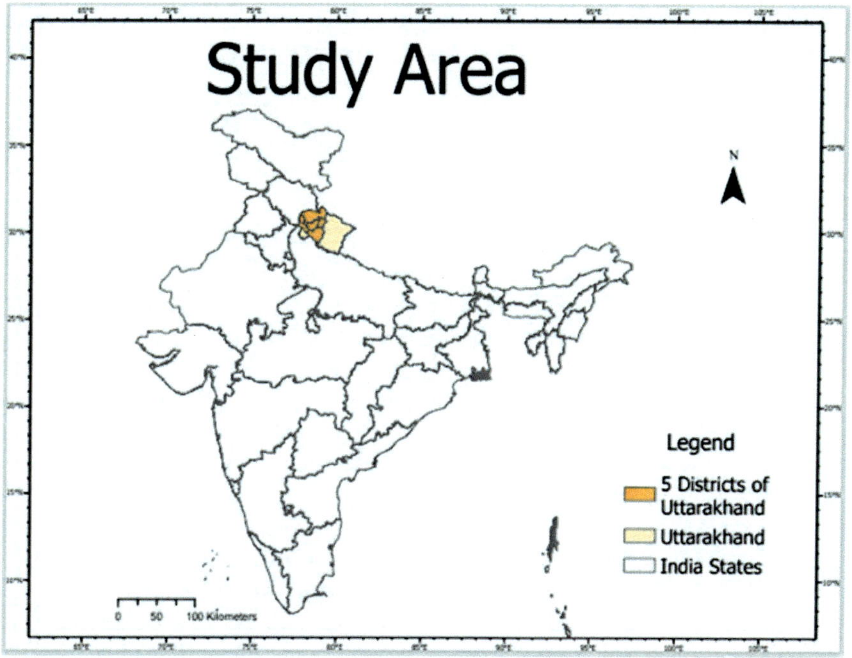

Fig. 2 Study area

3 Data Collection and Study Area

The study area is located in the northwest of Uttrakhand, lying in the 44th UTM zone and having a geographical extent of 30° 59′ 11″ N – 30° 34′ 15″ N and 78° 15′ 22″ E – 78° 44′ 24″ E Five districts selected for the study are shown in Fig. 2 and Table 1 shows the sources and images used for data collection.

Table 1 Data collection – details of the satellite images used for the study

Sl#	Data	Source	Type of image	No of images	Time period	Path	Row
1	Elevation	SRTM	1 arc-second (30 m)	9	May 2014	29	77
2	Slope	SRTM	1 arc-second (30 m)	9	May 2014	30	78
3	LULC	Landsat 8	OLI/TM Cloud Cover – 0–10%	4	May 2013	146	38
4	NDVI				May 2013	146	39
5	NDBI				May 2021	146	38
6	NDBI				May 2021	146	39

4 Results

Images were collected for May 2013 and 2021 and had approximately similar climatic conditions with 0–10% cloud cover to have clear photos. Landsat multispectral images are used for LULC and spectral indices evaluation. All the parameters used for the study are shown in maps, which help to have a spatial level understanding of the earth's surface at the time of the study. The study helped determine the behavior of the multiple parameters under consideration. The result section shows the study maps for periods 2013 and 2021.

4.1 Slope and Elevation

The slope of the study area ranges from 0 to 83.11°, and the elevation ranges from 253 to 7109 masl (meters above sea level), and the maps indicate that it is a hilly region with varying slope and elevation.

4.2 Land Use Land Cover (LULC)

The LULC map in Fig. 3 provides a visual interpretation of the changes in barren, built-up (development), cropland, forest, water, and snow in 2013 and 2021. The LULC changed in the area of the six classes in the years 2013 and 2021. As per the chart and table, it is clear that barren land increased in size and forest decreased by 14.94%. Snow and water changes were not considered for this study.

4.3 Normalized Difference Vegetation Index (NDVI)

NDVI shows the change in vegetation of the study area, and as per the map in Fig. 4 and Table 2, the low vegetation area increased in 2021 compared to 2013. It is evident from the map that the increase in low vegetation trend is towards the central part of the state where development is increasing. The map interpretation tells about the vegetation changes, and this spectral index denotes the form of land in both

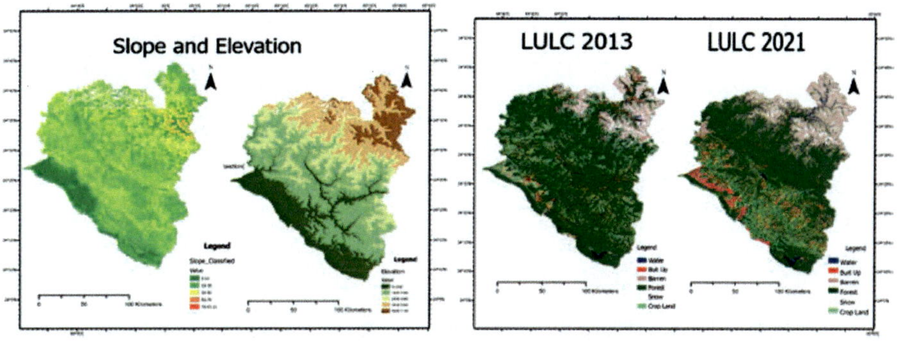

Fig. 3 Slope, elevation, and LULC maps

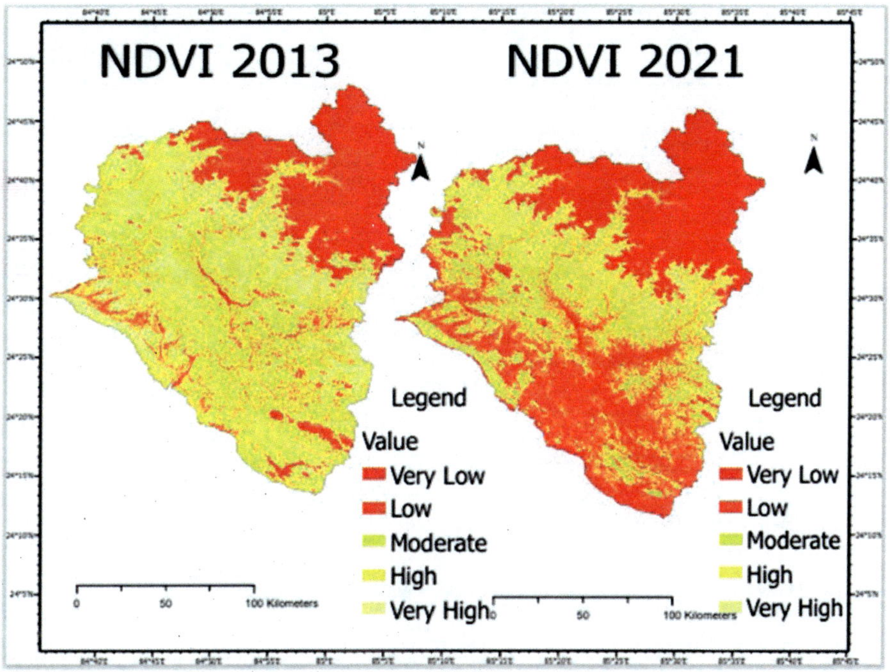

Fig. 4 Normalized difference vegetation index (NDVI) maps

years. High and very high vegetation reduced to 42.21% and 78.64%, respectively and moderately dense forests reduced in area, which denotes the reduction in forest land. Further analysis shows that low vegetation increased by 188.21%, and it can be assumed that moderate and dense forests were converted to cropland. Also, very low vegetation increased by 18.27%. This denotes the conversion of land to development.

Table 2 Normalized difference vegetation index (NDVI) changes between 2013 and 2021

NDVI	2013	2021	Percentage change
Very low	365.53	432.33	Increased 18.27
Low	184.63	532.13	Increased 188.21
Moderate	657.6	686.83	Increased 4.44
High	491.52	284.04	Decreased 42.21
Very high	299.3	63.92	Decreased 78.64

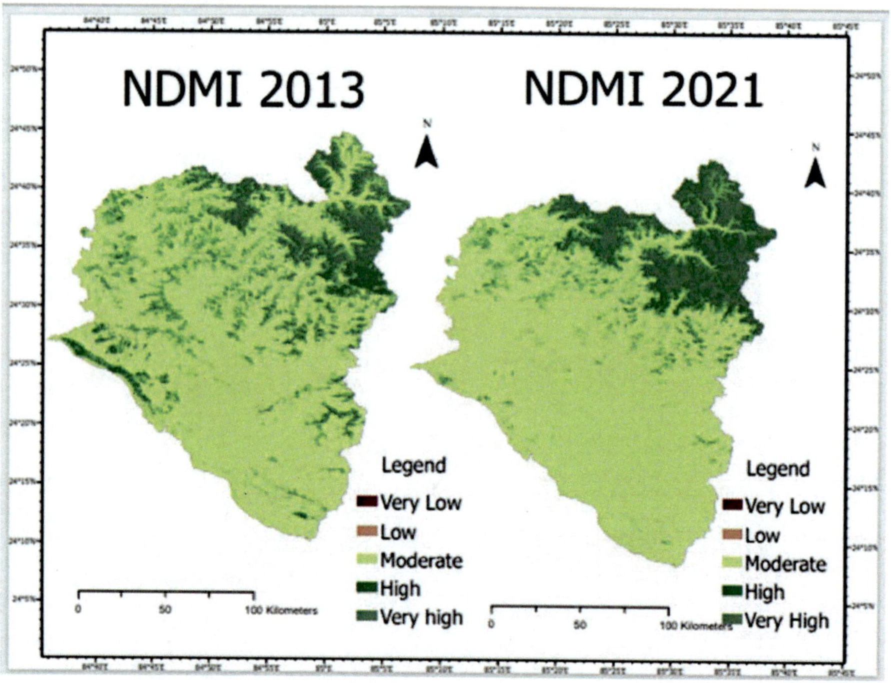

Fig. 5 Normalized difference moisture index (NDMI) maps of 2013 and 2021

4.4 Normalized Difference Moisture Index (NDMI)

The NDMI map in Fig. 5 and Table 3 shows the change in moisture index, and it shows that very high and very low values changed drastically. This can be an indicator of soil stability, and further study is required to enforce it. The table shows the values in 2013 and 2021 and the change.

4.5 Normalized Difference Built-Up Index (NDBI)

NDBI is the measure of built area, and it helps to map and evaluate the change over the years 2013 and 2021. As per the map in Fig. 6 and Table 4, the exceptionally high intensity built-area index increased by 29.56%, and it shows an increase in settlements/built-ups in the year 2021. Very Low and Moderate built-up index

Table 3 Normalized difference moisture index (NDMI) changes between 2013 and 2021

NDMI	2013	2021	Percentage change
Very low	0.0012	0.0001	Increased 18.27
Low	0.0187	0.0081	Increased 188.21
Moderate	1607.39	1602.43	Increased 4.44
High	308.02	219.49	Decreased 42.21
Very high	83.14	177.33	Decreased 78.64

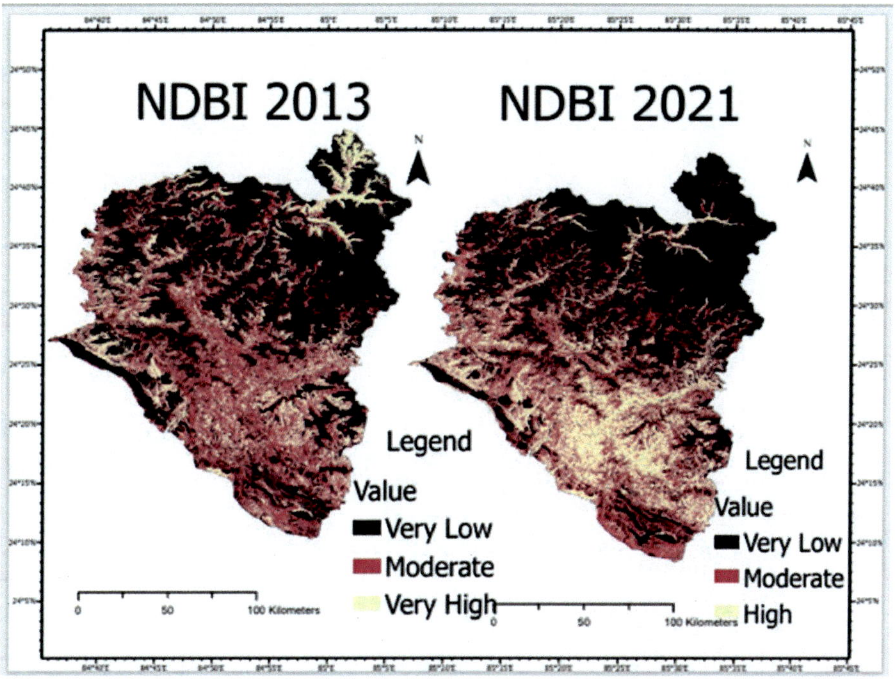

Fig. 6 Normalized difference built-up index (NDBI) maps

declined by 2.86% and 10.18%, respectively and it can denote the conversion of forest or cropland to barren.

4.6 Weighted Overlay

The parameters slope, elevation, LULC, NDVI, NDBI, and NDMI are reclassified in terms of degradation by assigning weight to each layer and in each layer as relevant. As the slopes become steeper, the ranks are assigned in increasing order, with the steepest slope being 5. The reason behind it is that high slope areas are vulnerable to degradation. The reclassification process is done from 1 to 5 for elevation, with the low elevation assigned to high value since low areas are susceptible to high degradation, and high elevation areas are mostly not preferred for development

Table 4 Normalized difference built-up index (NDBI) changes between 2013 and 2021

NDBI	2013	2021	Percentage change
Very low	866.7	841.91	Decreased 2.86
Moderate	777.92	698.72	Decreased 10.18
Very high	353.96	458.61	Increased 29.56

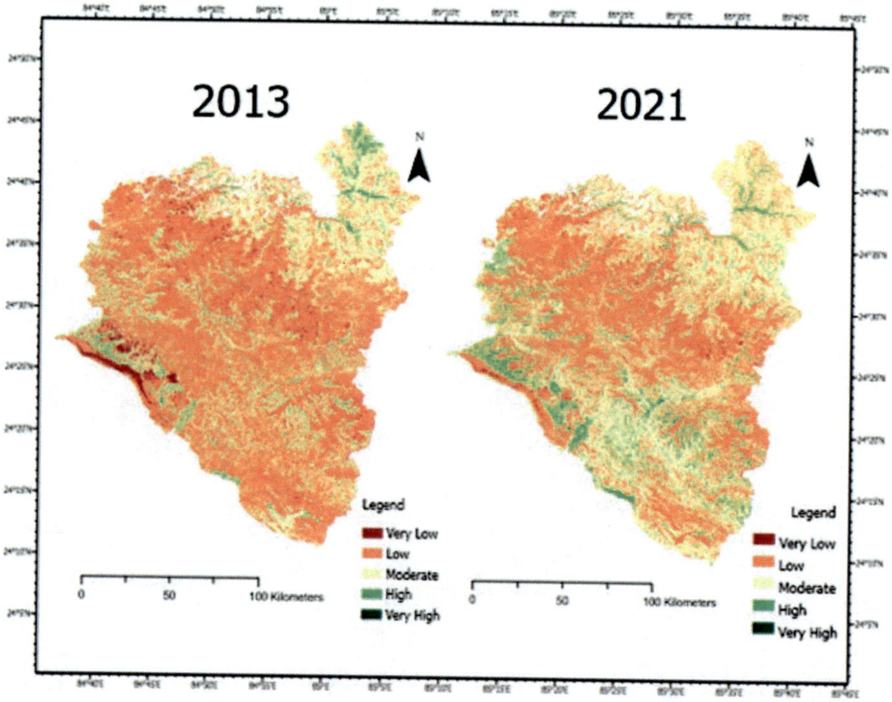

Fig. 7 Weighted overlay map showing degraded areas

activities. LULC is assigned with a high value assigned to built-up, followed by barren, cropland, and forest, and zero ranks are assigned to water and snow. NDVI is reclassified with a high rank assigned to less vegetated areas, and a low rank for more vegetated areas. For NDBI, a high rank is given for areas with more built-ups, and a low rank is provided for areas with fewer built-ups. For NDMI, low moisture areas are assigned less rank, and high moisture areas are in high rank. The second step is multiplying the cell values of the input raster by the assigned weight. Weights are assigned in the order of LULC to highest followed by NDVI, and NDBI. NDVI and NDBI are assigned equal weight followed by NDMI, slope, and elevation. The third step is adding the resulting cell value to get the output raster.

The weighted overlay map in Fig. 7 shows the overlay maps in 2013 and 2021 and is used to evaluate the degree of degradation during 2013 and 2021. The degradation analysis with spectral indices and weighted overlay methodology clearly shows the degraded areas visually and quantitatively. Table 5 shows the extent of change, and it is evident that degradation is more for high-ranked values.

Table 5 Weighted overlay table showing the change of degradation

Degradation intensity	Percentage change
Very low	Decreased 75.55
Low	Decreased 21.72
Moderate	Increased 49.89
High	Increased 56.1
Very high	Increased 96.38

5 Discussion

LULC, NDVI, NDBI, NDMI, slope, and elevation maps were used to analyze the most degraded areas. Degradation analysis is done by giving weight to each layer, and in each layer, the parameters are assigned ranks per the importance. In the case of NDVI, a high level is given for less vegetated areas, and a low rank is shown for more vegetated areas. For NDBI, a high rank is provided for areas with more built-ups, and a low rank is given for areas with fewer built-ups. In the NDMI map, low moisture areas are assigned less rank, and more moisture areas are in high rank. The slope is ranked from low to high since high slope areas are vulnerable to degradation. In the case of elevation, rank is assigned from high to low since low areas are susceptible to high degradation and high elevation areas are mostly not preferred for development activities. The weighted overlay is done with LULC, NDVI, NDBI, NDMI, slope, and elevation layers, and the resultant overlay map shows the degraded areas in low to high order (Fig. 5). Agroforestry degradation analysis is the first step in restoration services implementation. Identifying the parameters and the weighted matrix and ranking help to devise a suitable methodology to revive the ecosystem which helps to reduce adverse environmental effects. The study can be extended to evaluate the effectiveness of ecosystem restoration service, and EO can be used to calibrate and validate existing or new ecosystem service models. The study can be extended to evaluate other ecosystems' degradation extent and implement restoration methodologies.

5.1 Limitations

Apart from the parameters considered for the current study, there are multiple parameters that affect agroforestry degradation. Some of them are precipitation, soil erosion, and fluctuating population, which add to the intensity of agroforestry degradation. There were multiple disasters also in the past, such as earthquakes, landslides, flash floods, and forest fires in the study areas, which are not considered in the study. Analysis of the causes of disasters and the dependent parameters of agroforestry ecosystem degradation helps implement corrective measures to restore the ecosystem and mitigate the effects of disasters to some extent. Tree intervention in crop fields without the reduction in crop yield is the primary factor in the adoption of agroforestry [13].

6 Conclusion

Spatiotemporal analysis of the agroforestry ecosystem helps evaluate the degree of degradation and determine the parameters that accelerate the process. The study emphasizes the efficiency of remote-sensing data and GIS integration to assess the extent of degradation and uses spectral indices, a potent tool to perform degradation analysis in complex remote areas. By the selection of suitable parameters, degraded areas can be delineated to implement restoration solutions. Remote sensing with spectral indices can give more accurate results without in situ evaluation, and the scope is very high and can be extended for multiple ecosystem degradation analysis scenarios. Analysis of the causes of disasters and the dependent parameters of agroforestry ecosystem degradation helps implement corrective measures to restore the ecosystem and mitigate the effects of disasters to some extent.

Acknowledgments The paper is part of the senior author's MTech Thesis on agroforestry ecosystem degradation analysis and ecosystem restoration services. The senior author thanks God almighty for the abundant grace and mercy, my family for the endless support, and teachers for the guidance and support to complete the study. Also, the senior author expresses deepest gratitude to our Beloved Chancellor, Sri Mata Amritanandhamayi Devi, for providing a friendly environment to work in.

References

1. L. T. Luppino, F. M. Bianchi, G. Moser, and S. N. Anfinsen, "Unsupervised image regression for heterogeneous change detection," arXiv preprint arXiv:1909.05948, 2019.
2. R. Dhanai, R. Negi, S. Singh, and M. Parmar, "The effects of climate change on natural resources and socio-economic condition of Himalayan communities of Uttarakhand, India," *International Journal of* Modern *Communication Technologies and Research (IJMCTR)*, vol. 2, no. 11, pp. 21–25, 2014.
3. M. V. Ramesh, "Design, development, and deployment of a wireless sensor network for detection of landslides," *Ad Hoc Networks*, vol. 13, pp. 2–18, 2014.
4. S. Kumar and J. Srivastava, "An explained typology of natural disasters in the state of uttarakhand, india: Preparedness and response mechanism," *Environ Pollut Climate Change*, vol. 5, p. 242, 2021.
5. R. P. Yadav, J. K. Bisht, V. S. Meena, and M. Choudhary, "Sustainable agroecosystems for livelihood security in indian himalaya," *DeOliveira AB, edtor. Sustainability of Agroecosystem. IntechOpen: In*, pp. 63–77, 2018.
6. S. Mahato, S. Dasgupta, N. Todaria, and V. Singh, "Agroforestry mapping and characterization in four districts of garhwal himalaya," *Energy, Ecology and Environment*, vol. 1, no. 2, pp. 86–97, 2016.
7. I. Dev, A. Ram, S. Bhaskar, and O. Chaturvedi, "Role of agroforestry in current scenario," *Agroforestry for Climate Resilience and Rural Livelihood*, p. 1, 2019.
8. N. Vasudevan, K. Ramanathan, and T. Syali, "Land degradation in the western ghats: The case of the kavalappara landslide in kerala, india," in *F-EIR Conference on Environment Concerns and its Remediation*, pp. 199–207, Springer, 2021.
9. R. Mamtha, N. Jasmine, and P. Geetha, "Analysis of deforestation and land use changes in kotagiri taluk of nilgiris district," *Indian Journal of Science and Technology*, vol. 9, no. 44, pp. 1–4, 2016.

10. D. S. Varma, K. Nandanan, V. R. PC, B. Soundharajan, M. L. P'erez, K. Sidharth, and M. V. Ramesh, "Participatory design approach to address water crisis in the village of karkatta, jharkhand, india," *Technological Forecasting and Social Change*, vol. 172, p. 121002, 2021.
11. G. Prasad and M. V. Ramesh, "Spatio-temporal analysis of land use/land cover changes in an ecologically fragile area—alappuzha district, southern kerala, india," *Natural Resources Research*, vol. 28, no. 1, pp. 31–42, 2019.
12. S. Radhakrishnan and P. Geetha, "Urban sprawl assessment using remote sensing and gis techniques: A case study of ernakulam district," in *Intelligent Sustainable Systems*, pp. 293–307, Springer, 2022.
13. A. Bijalwan, P. Verma, M. J. Dobriyal, A. Patil, T. K. Thakur, and C. Sharma, "Trends and insights of agroforestry practices in madhya pradesh, india," Current Science, vol. 117, no. 4, p. 597, 2019.

Part III

Urban Planning, Pro-environmental Behaviour and Sustainable Development

Wood as a Catalyst for Construction and Architecture that Facilitates the Creation of Sustainable and Resilient Communities. The Case of the Palafitos de Chiloé, Chile

Stefania Pareti, David Flores, Loreto Rudolph, and Martina Pareti

Abstract

The present study explores how wood can act as a catalyst for construction and architecture that facilitates the creation of sustainable and resilient communities, preventing landscape degradation and restoration. We selected the stilt houses of Chiloé Island as a case study due to (1) the unique vernacular architecture; (2) the geographical morphology that has allowed from its genesis as a city to build in wood and sustainable materials; (3) the unique transportation system (minga), (4) management of its gentrification, despite being an iconic tourist destination in Chile, and (5) being a sustainable territory, with the management of its density, despite being an island and highly populated. We developed the methodology through the constructive analysis of a typical stilt house, woods used in its process, and interviews that show these constructions' internal dynamics. We conclude that wood acts as a catalyst for construction and architecture that facilitate

S. Pareti (✉)
Universidad Andres Bello, Santiago, Chile
e-mail: s.pareti@uandresbello.edu

D. Flores
Pontificia Universidad Catolica de Chile, Santiago, Chile
e-mail: daflores2@uc.cl

L. Rudolph
Universidad Tecnica Federico Santa Maria, Valparaíso, Chile
e-mail: loreto.rudolph@sansano.usm.cl

M. Pareti
Pontificia Universidad Católica de Valparaíso, Viña del Mar, Chile

J. T. Anderson (ed.), *Advances in Environmental Sustainability*, Springer Proceedings in Earth and Environmental Sciences,
https://doi.org/10.1007/978-3-031-26365-1_7

the creation of sustainable and resilient communities, preventing landscape degradation and restoration, as has occurred in the Chilote case.

Keywords

Wood construction · Vernacular architecture · Sustainable communities · Resilient communities

1 Introduction

Architecture and wood are essential components to form a sustainable vernacular style in the selected case study, Chiloé: An island in southern Chile surrounded by the Pacific Ocean (Fig. 1), specifically in the X Region of Los Lagos, between parallels 42 and 43 of south latitude. The geographical understanding of the island is fundamental because it characterizes the heritage value of the island from an architectural, cultural, and traditional point of view.

The arrival of eleventh-century Jesuits and ninth-century settlers was decisive in generating the island's architecture today. The Jesuit influence and the tenants' knowledge and cultural values maximized natural resources use. Wood is a fundamental material because the construction of the Chiloe culture has historically been based on it. It was conditioning both for the work of the Chilotes and their mentality and social life: its wooden habitat, its wooden tools, and its wooden paths, such as the one that joined Castro and Ancud [1]. The above is implied for geographical reasons. Its location in the southern hemisphere allows the development of an exceptional forest with vegetation primarily composed of a variety of trees such as Ciruelillo, Avellano, Coigüe, Ciprés, Canelo, Alerce, Ulmo, Mañío, among others. Each one has properties that are used to construct houses on the island.

The first stilt houses appeared in the twentieth century due to migration from the countryside to the city. They were born in response to a city saturated with urban infrastructure. Part of the population located their homes on the edge of the sea and the mainland. Homes were characterized by being simple volumes with gabled roofs, with one façade facing the sea and the other facing the street.

The arrival of carpenters and furniture makers in the period in which the island was opened to trade, was a trigger for building the stilt houses. This is primarily due to the platform system, where the built volume works independently of the supporting platform, which is valuable due to seismic activity.

2 Theoretical Framework

The theoretical framework of this study focuses on reviewing Sustainable cities, Landscape degradation, and restoration, wood construction, palafito, Chiloé Chile, and sustainable tourism. Cities are becoming increasingly complex systems from a social, economic, and ecological point of view [2]. Smart cities are defined as those

Fig. 1 Map of Chiloé in Chile. (Own Elaboration)

cities capable of combining people, technology, and creativity to make them more efficient and sustainable. Chiloé is the case of the creation of a sustainable environment. This intelligent and ecological city stands out in terms of geographic diffusion [3], evidenced in the control of gentrification since, despite being a tourist destination in Chile, it has managed to manage population density.

Vulnerability to change is a characteristic that smart cities have, an essential factor is the degradation of the landscape in the territory, where its transformation is understood as a spatial and visual expression of ecosystems [4].

The architecture and construction in Chiloé applied to the case study, the stilt houses is undoubtedly constituted by the resources available in the area: the use of wood as the main material in construction. The woods used are selected according to the functions they must fulfill based on their characteristics, providing rigidity to the structure or thermal insulation. The most resistant woods for the structure, such as Ulmo or coigüe, whether in its carved or rough condition, for the Mañío and Las Guaitecas cypress windows. The same for the cladding together with Mañío or larch [5]. These woods are used in the construction of stilt houses, structures composed of driven pillars, and are born in response to context and need. They are located mainly in calm water bodies such as lakes, lagoons, or rivers. However, in the case of Chiloé, they are explicitly found in a sea edge condition and on land [6].

The protection at ground level against humidity distinguishes this typology, using the residual space as a woodshed or domestic warehouse [7], with a façade that faces the street, where there is a patio that serves as a vegetable garden or chicken coop. Another side faces the sea, where the residents' fishing boats are moored.

Currently, most of the stilt houses on the island of Chiloé remain standing due to their conservation for tourist purposes rather than housing. Tourism has grown as a component of the economy of the region. Tourism is promoted as a means of economic development. Simultaneously, concerns about sustainability due to the cultural, financial, and environmental damages that tourists can cause, notably pollution and overdevelopment [8].

More and more urban centers choose to implement a Smart City model. Although there are some cities much more advanced than others, such as Tokyo, where technologies that increase efficiency and respect for the environment are implemented, be it energy storage or the use of electric vehicles [9]. Toronto has planned and developed living spaces from scratch to achieve a community focused on technology, addressing problems of both urban expansion and climate change, accessibility, and also efficiency [10]. Cities not as advanced as Chiloé have gradually implemented measures to become Smart cities, such as signing agreements in collaboration to promote joint work for tasks related to protecting or conserving architectural heritage. Avoiding degradation of the landscape, culture, and environment [11] and encouraging restoration of damage caused by fires, pollution of the sea, or the misuse of available materials, stands out in the Amazon [12]. Cattle ranching and logging of forests are transforming humid tropical forests into barren landscapes and soils, so restoration of these requires understanding and respecting the cultural attributes of the landscape, beliefs, and values, as in the case of Australia

[13]. Australia shows environmental narratives, such as integrating biophysical and cultural aspects to restore the ecology. The vulnerability of Smart cities, together with the degradation of the landscape, is evident in the recent fire in Castro, Chiloé, on December 8, 2021 [14], where more than 5 hectares with approximately 120 homes were affected. Environmental narratives are implemented for its restoration, including the Regional Committee for Disaster Risk Management. Defining damage control strategies and determining actions to support the affected population and the islander's homes built primarily of wood [15].

The stilt houses that prevail in Chiloé have been treated and maintained correctly, have remained standing after fires or earthquakes, and have not been demolished due to a lack of compliance with health standards. Still, the number of stilt houses standing is reduced. These stilt houses, made of wood that lies in contact with the water, are from Ulmo. The cladding is made of larch wood. Both kinds of wood are extracted from the area, which denotes sustainable construction of these buildings. Similarly, the supporting structure for Villa Palafítica de Bad Buchau [16] is a series of bamboo wood piles driven into the water. Its resistance, flexibility, and durability make it a good material for stilt house construction. In the case of the Niyishi Communal House [17], although the wooden structure is not in contact with water, it can withstand climatic variations.

As mentioned above, the palatial structures still standing in Chiloé are the minority. They are mainly restaurants, boutique hotels, or other tourist attractions. This brings a series of problems related to the area's gentrification. The attention and admiration given to stilt houses, both nationally and internationally, has been positive for restoring these structures. The government has subsidized those who have lived in them for years. The intervention involves changing more than 50 posts that keep the palafito standing [18], cladding, and door and window frames.

Tourism on the island stands out not only for its natural landscapes and centenary architectural heritage but also for the sustainability offered by the accommodation services. This is primarily due to the community's certification and participation, the culture of self-consumption, and the products produced in the area [19].

3 Methodology

The study will be carried out mainly through analyzing secondary sources, such as writings and public access interviews with relevant actors for the case study. Especially architects who work in the area understand the internal dynamics, both social and constructive, and agents linked to the restorative processes of Chiloe architecture and administrative procedures of a fundamental cultural heritage.

First, it is necessary to understand the historical development of vernacular architecture that today is recognized as the imprint of an urban image associated with the Chiloé archipelago in close relationship with wood as a letter of introduction. This analysis is not raised clearly from a material perspective since a fundamental part of this style is associated with an idiosyncrasy typical of the archipelago.

In addition, understanding its geographical context as a determining factor of architectural development is key to carrying out any analysis in the area.

This first analysis will make it possible to relate materiality and geography to descend to the stilt houses in response to these and other conditions, achieving unique habitability in the border sea context. The analysis seeks to establish this typology as an expression of both material culture and an immaterial culture and, therefore, poses the object as worthy of protection, restoration, and promotion, which would be the second branch of analysis of the study.

Finally, the study raises the current condition of the palafito. First as part of Chiloé's architectural heritage, second as part of its cultural heritage, and third as a catalyst for development under the two conditions already mentioned. It is setting guidelines for restorative and interventionist processes that take the palafito as a basis for developing a new sustainable architecture based on wood. This analysis is not done only from the condition of a valuable object that the palafito possesses but is taken as a means to achieve the development of the associated communities, understanding the building as a generator of development through restoration, intervention, and promotion.

4 Results

Given the insularity that being an archipelago implies, the development of Chiloé is strongly conditioned by its geography [20]. Thus, three main macro zones account for this perspective: sea, land, and the sea's edge. The first is a limit to be traversed, the second is an available space, and the third is a diffuse limit between sea and land. This space for interaction becomes relevant given the tidal influence of the archipelago that constantly varies, blurring the end line between the two other macro zones. Thus, from a material perspective, the urban development of Chiloé was, is, and will be conditioned by its geography, which, despite implying restrictions and difficulties, gave way to the consolidation of a particular type of city that adapts to its context. The citizens know how to respond to this with the available resources, building a way of living worth observing [20].

The consolidated city, in turn, gave way to constructing a typical imaginary for the archipelago, an urban image where wood is painted as the protagonist and in response to the need to inhabit. This decision is far from random and is argued from the base cultural syncretism of the Chiloe idiosyncrasy, resulting from the interaction of native peoples and Jesuits who arrived in the area. One of the results was the creation of a unique carpentry school that was capable of responding to the needs of the archipelago. Thus, vernacular architecture was consolidated, based on wood, where simple volumes, low-rise constructions, and compositions that prioritize the full over the void predominate, all decisions mainly associated with the climatic characteristics of the sector and that, therefore, sought to achieve thermal comfort [20]. Historically, houses were built with the kitchen as the center, using a stove to heat the rest of the property.

The stilt houses were born based on these same construction precepts, added to factors such as country-city migration, the limited availability of land, and productive activities. Together with the churches and minor elements, such as larch tiles, are the icons of the Chilota architectural image.

The palafito as a typology maintains the aforementioned constructive logics, both material and spatial, but with a apparent particularity that makes it remarkable as an object of study of the vernacular Chilote; its relationship with the edge of the sea. These constructions are on the sea, supported by wooden piles from the shore. Thus, the palafito presents unique habitability within the archipelago, where the sea becomes the house's patio, constantly changing along side the ups and downs of tides through out the day.

Its origin and location, inscribed within the framework of the country-city migration of the early twentieth century [21], also have to do with the relationship between the chilote and the sea. Mainly due to the notion of a productive place and granted sustenance. Despite their current value, stilt houses were not always viewed favorably and conveyed a particular image of precariousness [20], given the self-constructing processes that gave rise to them. This added inclemencies such as fires and earthquakes, resulting in the loss of a large part of these buildings, and began to raise an awareness movement regarding the value of this typology, promoting the first signs of care and protection. Today this type of housing represents not only a way of life and typical daily life of the archipelago, but they are part of an island heritage that over the years has become more relevant, so much so that today its conservation and restoration are vital to the archipelago's development programs [21].

The current condition of these buildings is framed within the concept of living architecture [20, 21]. Understanding first the close relationship with the user population, that is, they are buildings in use, and second, highlighting the constant programmatic evolution of these and their adaptability to their context, being able to cope with the passing of the years and the new needs that this implies. The finite condition of the wood, understanding its susceptibility to elements such as fire, xylophages, and the same humidity of the environment associated with the constant presence of the sea, gives an approximate life of 60 years [21] to constructed buildings in wood. It is necessary to highlight the role of restoration and understand the cultural importance of the buildings that are the object of this study.

Under this concept are grouped three critical elements: heritage, sustainability, and resilience. They are identified as a line for the development of cities with a view to the future, where the focus is placed on communities and neighborhoods as the fundamental and primordial unit of the city. The three can also be observed with wood as a flagship material to carry out the process.

The relationship is not random but has to do with the current role of heritage within the city, both built and cultural. In this sense, the stilt house is not considered within this category only because of the current image it transmits (which is still valuable by itself for no reason). Rather its historical development, relationship with the chilote, what it means for the current habitability within the archipelago, and the identity roots that it represents both for its direct users and the rest of the archipelago as they are icons of its culture and idiosyncrasy. In this sense, current heritage discourses indicate that there is no heritage without a community that values it. Therefore, the protection, restoration, and promotion of origin, in turn, is the community's protection, restoration, and promotion. Thus, it advocates sustainable groups from a physical and immaterial perspective, which in turn means promoting resilient communities capable of adapting and surviving the cultural and urban changes of which the city is a part.

Thus, the restoration of the material object, in addition to filling deficiencies that today affect buildings, reinforces the cultural heritage associated with it, lengthening its useful life, understanding that these are the bodies where both a material and an immaterial culture fall. At the same time, its restoration influences the quality of life of those who inhabit the stilt houses. Therefore, it positively affects the habitability of these spaces; it works for the directly associated community.

In this sense, taking care of problems such as rotting or corrosion of exterior materials or solving wear and tear and lacking some elements are the main pathologies suffered by stilt houses [21]. Maintenance becomes a highly significant action for the urban and social development of the archipelago, also understanding the role of social catalysts that this type of building has today.

Related to the latter, another critical factor within the heritage logic that falls on stilt houses and especially Chiloé is the current role of tourism within the dynamics of the archipelago, given its transformation into a tourist attraction due to its geography, culture, and its architectural and landscape image. In this sense, the palafito becomes the central element due to its role within the imaginary chilote. Therefore, a process of reconversion is associated with these processes, where homes cease becoming hotels, shops, and restaurants, among others [20]. This programmatic reconversion has two crucial aspects influencing sustainable and resilient development issues in the urban structure and the associated communities. In the first place, to understand the current role of the latter within the conversion processes, understanding their importance for the valuation process of the palafito; the association of these with their direct community is what differentiates it between a truly patrimonial and valuable element and a museum object, only admirable. In other words, avoid gentrification and displacement of the original population; without chilote, the palafito is not valuable. And secondly, once the first point is resolved, the ability of these buildings to be reconverted to receive new programs, which in turn implies new ways of living, which suggests taking care of unique needs.

Thus, new materials, construction techniques, and technologies, in general, can meet the needs of these new programs to enter the dynamics of restoration and promotion. Therefore, it is worth asking how these elements are integrated to promote a living heritage without affecting its dynamics and the typological bases that make it heritage. Within the specificity, it is also worth asking how these new materials dialogue with wood considering its role within the imaginary chilote.

These questions are not new, given that this phenomenon associated with tourism and the reconversions it implies has happened for years. Therefore, finding satisfactory answers that take on the new roles that these buildings are adopting while respecting heritage values is not difficult. It is enough to see, for example, the work of the architect Edward Rojas [20] within the archipelago, among other exponents that follow this line, where there are also good examples within the typology of stilt houses, but also of the entire vernacular chilote.

Thus, the correct performance of the restoration, reconversion, and intervention works within the stilt houses would fulfill three fundamental premises if one wants to consider sustainable development in the future. From a material perspective, stilt houses are associated with a self-sufficient logic, given the use of wood as the primary construction material and the constructive sense of climatic comfort. Without going any further, sustainability concepts were discussed within architecture long before they took on the relevance they have today. Added to this, the adaptability of this material to new technologies that come to answer current problems works along the same lines, creating a base typology resistant enough to adapt to contemporary times in pursuit of its promotion towards the future.

From an urban and social perspective, the role of social and urban catalysts is reaffirmed (under a logic of centers), worth the redundancy, posing as development and economic foci for the entire city. Investing in these places brings monetary benefits to the city. It raises these places as meeting spaces and cultural interaction, which ultimately promotes the appreciation of the Chiloe idiosyncrasy and its transmission for both new generations and visitors. The stilt houses are associated with tourists, who, from the perspective of the experience of the place, become part of the dynamics, not to admire them as museum objects, but to become part of them as a new actor within the valuation of the place. Likewise, the process is an example for other structures with similar dynamics within the archipelago and where wood has the same or even more influence.

Finally, from a cultural perspective, one cannot speak of sustainable development if one does not consider the cultural richness of the places to be developed, understanding this as a fundamental part of the dynamics of a place. This condition is even more evident in a place like the Chiloé archipelago, where it is part of daily life. Therefore, its omission is a setback in the development processes, considering the role of the communities, as argued. The sustainable and resilient development of the archipelago is inextricably linked to its cultural heritage. Just as this is an essential part, there is no Chiloé without its cultural heritage.

5 Conclusion

Society cannot take a purely physical and material route to carry out an integral sustainable development, primarily focused on the community. This is only one edge of the problem. In this sense, the correct product associated with sustainability should take as logical material, social, and cultural guides. In this sense, the palafito is taken as an object of study both for its materiality, urban, and role within the Chiloe culture, presenting itself as a vital icon of the archipelago's imagination. Thus, it stands out for using wood associated with its low carbon footprint and thermal properties and its ability and adaptability to understand the new problems that contextual changes entail. These challenges include the arrival of tourism or the appearance of new technologies without affecting a typological base with the material's imprint. In addition, it stands out for its close relationship with the identity of the Chilotes, as it is the object of tradition and materialization of an ancestral culture fundamental to understanding the area.

Thus, we do not speak of only physical transformations regarding the aforementioned programmatic changes. Still, we discuss sustainable development based on resilient communities by maintaining an identity reflected in the stilt house. It is a cultural object, and we must understand its relationship with the sea, the seashore, and the land and what this means for the chilote. The stilt house can survive by adapting as required by the new needs of its context, giving way to new spatialities and allowing new social and urban relationships, but without altering its fundamental origin. Despite the changes, we are still facing the materialization of centuries of understanding of a specific physical and cultural context that gives way to a cultural richness full of popular knowledge and traditions, where its promotion and restoration means the transfer and survival of community logic fundamental, all this, under the eaves of wood as a famous material to build the city.

References

1. Barrientos H, s.f Iglesias y Capillas de Chiloé (Revista de Divulgación Centro Chilote, N°6, 7–10 pp.
2. Liu J, Dietz T, Carpenter S, Alberti R, Folke M and Moran C. Complexity of coupled human and natural systems. 2007
3. Cugurullo F Exposing smart cities and eco-cities: Frankenstein urbanism and the sustainability challenges of the experimental city
4. Romero M, Morera C, Alfaro D, Marín R and Barrantes O. Degradación del paisaje natural en territorios urbanos, medido a través de métricas del paisaje. 2013
5. Pastor E. El bordemar en la isla de Chiloé. 2018
6. Piesik S. Hábitat: Vernacular Architecture for a Changing Planet. 2017
7. García D. Chiloé: Reciclaje arquitectónico. 2020
8. Culver Lab. Cultural tourism and sustainable local development. 2011
9. Yamagata Y, Simulating a future smart city: An integrated land use-energy model. 2013
10. Tierney T F, Big Data, Big Rhetoric in Toronto's Smart City. 2019
11. Servicio Nacional del Patrimonio Cultural. Dibam y Fundación Amigos de las Iglesias de Chiloé firman convenio de colaboración. 2015
12. Nepstad D, Uhl C and Serrao E, Recuperation of a Degraded Amazonian Landscape: Forest Recovery and Agricultural Restoration. 2014
13. Robertson M, Nichols P, Horwitz P, Bradby K and MacKintosh D, s.f. Environmental Narratives and the Need for Multiple Perspectives to Restore Degraded Landscapes in Australia, 2000
14. Teletrece, [Online] https://www.13.cl/programas/tu-dia/noticias/tragedia-en-castro-devastador-incendio-consumio-mas-de-140-casas Ultima visita 20-12-2021. 2021
15. Guerra I, Incendio en Castro: Elevan a más de 100 las viviendas afectadas por las llamas y a 15 las personas lesionadas. El Mercurio. 2021
16. Maier U and Harwath A. Detecting intra-site patterns with systematic sampling strategies. Archaeobotanical grid sampling of the lakeshore settlement Bad Buchau-Torwiesen II, southwest Germany. 2011
17. Teletrece, Obras de "rescate" de los palafitos de Chiloé llegan al 75%. Video on line: Last visit 20-12-2021 (https://www.youtube.com/watch?v=_3Rzl3iBMMg&t=57s). 2014
18. Agrandoña G, Chiloé y Huilo Huilo, destinos sostenibles. 2019

19. Rojas, E. Artifica tu casa 2021: Edward Rojas habla sobre la arquitectura de Chiloé, [En línea]. Available:https://www.youtube.com/watch?v=WY2ZLPGSCdg. [Last visit: 2021 december 19] 2021
20. Ministerio de Vivienda y Urbanismo. Guía de recomendaciones para la reparación y puesta en valor Palafitos de Castro. Los Lagos, Chile. 2013
21. La Estrella de Chiloé (23 de Febrero 2020). Capacitan a comunidades de la provincia para que protejan a sus iglesias p. 13 https://www.laestrellachiloe.cl/impresa/2020/02/23/papel/

Identifying the Most Significant Factors Affecting Urban Housing Sustainability and Their Scales/Sectors of Influence: A Systematic Review of the Recent Literature

Mohsen Sanei, Mina Khodadad, and Fernando Calonge Reillo

Abstract

Housing is always crucial for the sustainable development of communities, specifically in urban areas, due to the population density of cities. The present study constructs its own structure based on the recent papers investigating various sustainability factors for the urban housing sector. A comprehensive systematic literature review gathered one of the most extensive lists of urban housing sustainability factors from 118 recent related papers. The factors are prioritized by their frequency of investigation and categorized by their scale(s) and sector(s) of influence. According to the results, the top three significant factors affecting urban housing sustainability are "natural resource or energy consumption/efficiency of the building/equipment (during the construction, operation, etc.)," "materials performance (durability, cost, thermal capacity, permeability, ability to re-use, recycled, eco-friendly materials)," and "access to public services/infrastructure: availability/quality of services and/or distance/time of travel time to the services (public transport, education/health/shopping/leisure facilities, parks, etc.)." By analyzing the results with an integrative approach, we found that environmental factors are the most considered ones, whereas institutional factors received the least attention from sustainability sectors. Also, the most significant measures are the ones that have impacts on both the 'building' and 'neighborhood/community' scales. We noticed that the neighborhood/community scale

M. Sanei (✉) · F. Calonge Reillo
University of Guadalajara, Guadalajara, Jalisco, Mexico
e-mail: mohsen.sanei@alumnos.udg.mx

M. Khodadad
Tecnológico de Monterrey, Monterrey, Nuevo León, Mexico

J. T. Anderson (ed.), *Advances in Environmental Sustainability*, Springer Proceedings in Earth and Environmental Sciences,
https://doi.org/10.1007/978-3-031-26365-1_8

indicators are almost as important as the measures that affect the building itself. The results of this study can help establish future housing-related policies and more efficient housing sustainability assessment tools.

Keywords

Sustainability assessment · Sustainability indicator · Urban housing sustainability

1 Introduction

Houses are essential factors shaping human communities. No more can they be considered as simple roofs to cover people. They also produce employment, create wealth, mitigate natural disasters, and help societies achieve integrated physical, economic, and sustainable development [1]. According to UN-Habitat (2012), besides other factors, the location of houses, their design and construction procedures, and the quality of their integration with the existing social, environmental, cultural, and economic sections of communities affect the people's daily lives, safety, health, and welfare [2]. This points out that housing is a central notion of sustainable development. The concept of sustainable housing not only indicates zero-energy, low-carbon, green, or high-performance houses but also includes the features of resource usage, energy efficiency, socio-cultural and natural structures, financial requirements, and people's lifestyles [3–5]. Some of the 17 Sustainable Development Goals and their 169 supplementary targets that present massive benefits and chances for the community, businesses, and the ecosystem are specifically relevant to the housing sector, focusing on society, well-being, and place-making [6]. The necessities of sustainable development in cities are becoming increasingly critical [7], and it is essential to come up with some solutions for spaces that can increase the social capacity of the community [8]. Consequently, the new global development emphasis on urban areas goes further than the usual concentration on slum upgrading and housing to comprise secure, accessible, affordable, and sustainable transportation, integrated and participatory planning, public and green spaces, enhanced waste management and air quality, natural disaster risk reduction and climate resiliency [9].

A significant part of societal failures is connected to housing in one way or another. Throughout the world, housing forms societies, repentantly on many occasions, by generating inequalities and disintegration. The contrasts between slums and gated neighborhoods, homelessness and vacant homes, overcrowding and sprawl, and other similar examples can help understand these inequalities and fragmentations. On the other hand, the resulting models end up with environmental, social, and economic expenses much higher than what most societies can grant. Therefore, the future of urban development will be based on how communities prioritize housing in their public argument regarding sustainable development [10].

The UN-Habitat has expanded attempts to reposition housing as a priority in arguments about sustainable urbanization. The rising necessity to deliver adequate housing to many people and the requisite to ensure a sustainable prospect for communities require a paradigm shift in housing-related practice and policy. Stimulating well-founded housing policies is vital in other related sectors, such as climate change, resilience, energy consumption, and mobility [11]. As sustainability needs to be defined and used as a tool to uphold policy and decision-making procedures, it should be supervised, assessed, and reported. Consequently, enhancing the management of international, national, regional, and local policies, strategies, plans, and projects is crucial to reaching more sustainable results with fewer adverse impacts on humans and nature [12]. Here, sustainability indicators can be of great importance. They are not only standard performance metrics but also imperative tools to support urban development [13].

According to Mateus and Bragança (2011), developing and using sustainability measurement techniques and building benchmarking are possible approaches to reaching a more sustainable built environment [14]. Consequently, sustainability assessment of houses can play a significant role in the sustainable development of cities, and to do so, there should be a collection of indicators [15]. This is why there has been an enlarged concern about developing and utilizing sustainability indicators for assessing the influences of housing development projects [16]. Providing the indicators linked to sustainable urban housing is critical for an effective, sustainable housing policy [17]. Academics and experts have developed several sustainability evaluation methods and indicators for construction projects, some of which look after the housing sector. However, it is still a question in theory and practice what factors should be more prioritized among others. The present study primarily tries to answer this gap of knowledge.

According to Mazon et al. (2018), it is logical to presume that particular sets of sustainability indicators should be present at different scales, including universal, national, regional, and institutional levels [13]. Moreover, various sustainability indicators and assessment systems are employed to evaluate sustainability in housing-related projects in multiple countries [16]. This study gathered factors from the literature from different areas of the world (with possible differences in indicator collections and assessment methods) at a universal scale. By doing a comprehensive systematic review of the recent literature in the field, this study tries to deliver an inclusive and compelling collection of sustainability factors that are influencing not only the housing itself but also its surrounding community. The foundation of our research is gathering, prioritizing, and analyzing the factors affecting the sustainability of urban houses (we do not consider the factors that are primarily related to the sustainability of rural housing). However, we include the factors influencing both urban and rural sectors in the results. Each factor can comprise multiple sustainability indicators, as so many indicators related to urban housing sustainability are studied in the literature. Also, we have tried to investigate the scale (building and neighborhood/community) and sector (different sustainability dimensions) of influence for each factor. This has resulted in an integrative analysis of the scale and sector of impact to clarify the importance of each one in the overall sustainability of houses and to understand which are primarily considered in the literature.

Accordingly, this research has constructed an inclusive prioritized set of factors (based on the frequency of usage in literature) and their scale(s) and sector(s) of influence. Since this study is a part of a project that aims to assess the priorities used in international Sustainability Assessment Tools (SAT; e.g., LEED [18, 19], BREEAM [20, 21]), completely behavioral indicators (e.g., consumption behaviors) are not considered in the results, as these factors are not considered in most housing-related SATs. The results can fill the knowledge gap about the most significant factors affecting the sustainability of houses (on a global scale) and guide policymakers to make their decisions and strategies more precisely regarding the needs of the housing sector. The outcomes, as well, can give designers, planners, managers, and other related authorities a deeper understanding of issues considering housing sustainability and can help evaluate currently used SATs for the housing sector.

1.1 The Considered Dimensions of Sustainability

It is not possible to make an exact definition of sustainability in architecture [22]. However, the objectives of sustainable development are originally defined for the economic, social, and environmental dimensions. These are the three well-known sustainability sectors, also known as pillars of sustainability, that have been studied for decades. Some authors, such as Chiu (2004), have studied another dimension, including cultural aspects [3]. Furthermore, for efficient compliance and considering sustainability features, like justice, the sustainability sections should be accompanied by an institutional dimension. Consequently, for this paper, sustainability comprises four dimensions: socio-cultural, economic, environmental, and institutional. The reason for merging the social and cultural issues into one dimension is the strong two-sided connection [23], the closeness of the concepts, and having common aspects [3].

This study defines the environmental dimension as any bio-geological procedure and its features. The socio-cultural dimension consists of the personal potentials of humans and their surroundings, such as skills, values, aesthetics, norms, interactions, and practices. The economic dimension includes the formal economy and any informal action that delivers services to people (individual/group), escalating the standard of living beyond the financial aspects [24]. Also, the institutional dimension contains issues regarding regulations, institutional cooperation, policy-making procedures, standards, and other similar elements.

2 Methodology

We conducted a comprehensive systematic review of the recent related literature. For this purpose, we used the Scopus search engine. Some keywords were selected according to the research topic to find relevant results for the review. The query structure began with searching the items "sustainability OR sustainable" and "factors OR indicators OR measures" in TITLE-ABS-KEY fields and the items "house OR housing OR home" in the TITLE field (Fig. 1). The document types were limited to "article," "conference paper," "review," and "book chapter," which were in the "final" stage of publication. For the publication year, the last recent years from 2015 to 2020 (including 2020) were chosen as we wanted to investigate the contemporary approaches used in the studies. As subject areas, "engineering," "social sciences," "environmental science," "energy," "business, management and accounting," "economics, econometrics and finance," "decision sciences," "arts and humanities," and "multidisciplinary" were selected to provide a great range of relevant results for each sustainability sector. Finally, "English" was marked as the language of the publications. Overall, we found 512 results. Then, through three stages of title, abstract, and paper reviews, we dismissed 394 results and selected 118 papers for the final evaluation. The dismissed papers were neither related to the purpose of this article nor appropriate in terms of the study scale (e.g., too general or detailed in subjects, such as investigating the solar panel specifications).

We reviewed the selected literature carefully. The whole text (not only the results and discussion parts) of the papers was read to find the factors influencing the sustainability of urban houses. This means that although the reviewed literature has been published recently, many factors originally from other older papers were also extracted from this literature. Therefore, the gathered indicators are not only based on the literature published since 2015. It should be noted that the factors affecting only some specific types of houses, such as log houses or emergency ones, were not marked. However, we included those factors that could apply to the general urban housing sector. In selecting the indicators, we also excluded the elements which were too broad or limited in definitions, such as quality of life or the material albedo coefficient.

As there is a vast domain of factors and indicators affecting the sustainability of urban houses, we were obliged to simplify and brand many of them in classes as the final factors. We tried to categorize the elements simply and understandably. The attempt was to classify all related aspects to one issue, as much as possible, into a separate category. As categories and factors are highly related in some cases, some measures might be classified in more than one category. For example, the cost of materials is a measure that is considered in F2: material performance, F51: construction cost, and F62: renovation/repair/reconstruction cost. According to the authors' opinions, if some measures are considered unique or noticeable, they are

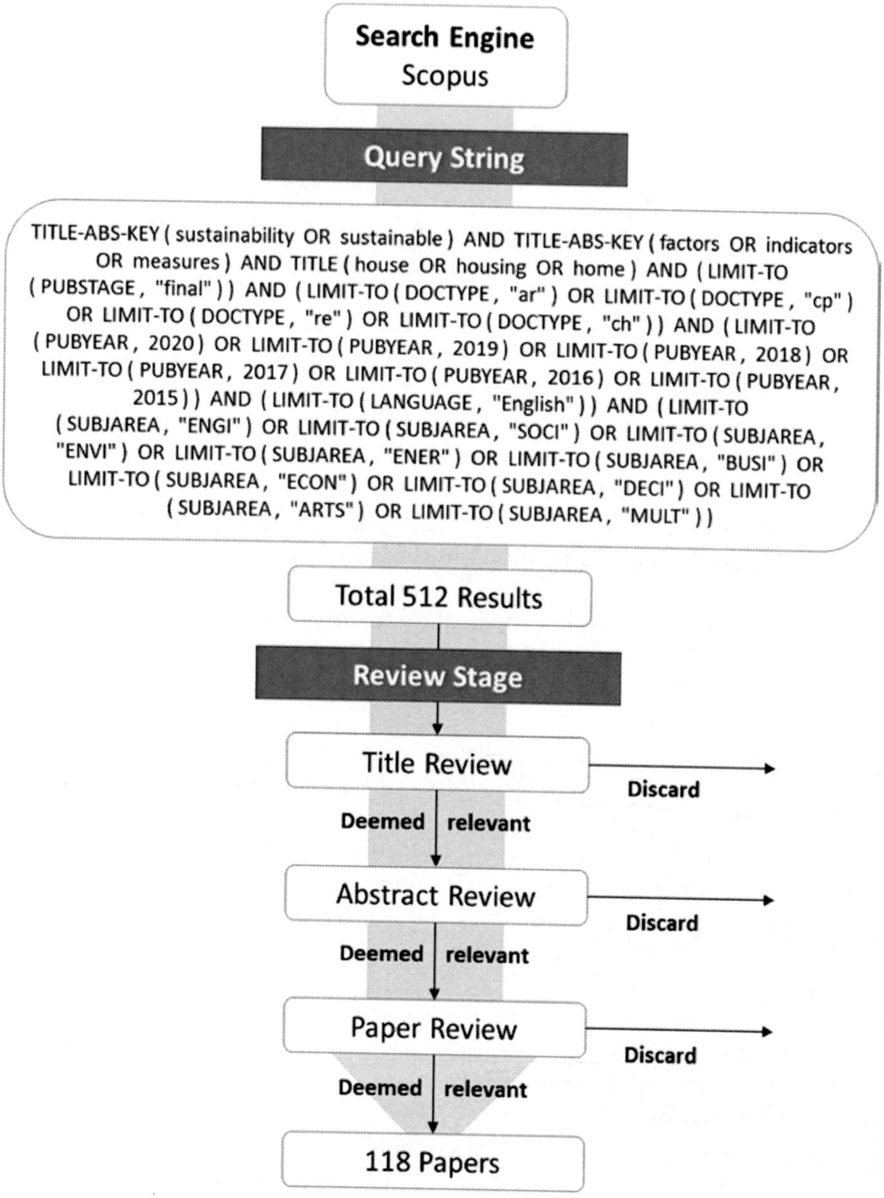

Fig. 1 The structure chart for the literature selection process

listed separately. As mentioned before, we excluded entirely behavioral factors (considering their direct effects). Also, we considered the factors that are only partially related to housing but are essential in the overall sustainability of the housing sector (e.g., affordable housing costs).

Additionally, only the factors with the local scale of influence (building or neighborhood/community) were collected from the papers. We did not consider others with bigger scales of influence (e.g., regional/national scale) as they do not directly affect buildings. The exceptions to this are some institutional factors, such as building codes and standards, which can directly affect the local housing criteria. Finally, we included the measures that only partially play roles in housing sustainability, but their roles are crucial. An example of these measures could be the affordability of housing-related costs (purchase price, rent price, etc.), which are partially related to the house and the household income rate. Therefore, it is vital to notice not all these factors may help measure the actual performance of buildings. However, these data are crucial for decision-makers in the housing sector to know what elements and subjects require more attention regarding overall housing sustainability.

The final result is a prioritized list of factors gathered from the papers. We analyzed the terms based on investigation frequency (IF, number of repetitions in the literature for each factor), investigation coefficient (IC, investigation frequency divided by the total number of the reviewed literature), the scale of influence (building and neighborhood/community), and direct impacts on sustainability sections (environmental, socio-cultural, economic, and institutional dimensions). In this research, we scrutinized each factor's frequency to show how often a factor is investigated and, therefore, to be a base to calculate the degree of importance that the literature has given to each scale/sector of sustainability. The qualitative analysis regarding the recognition of the scale and the area of impact for each factor was done based on the authors' knowledge, not on the findings from the reviewed literature. Each factor may influence multiple sectors or scales.

3 Results and Discussion

Papers were distributed worldwide based on the authors' address affiliations (Fig. 2).

As depicted, the results of this study are not based on the priorities of a limited region in the world, but it is constructed in a more or less universal context. The critical issue here is primarily the development level of countries. It can be seen that there are examples from the developed, developing, and also the least developed countries among the results, and there is not any significant consideration of one part among others. Therefore, the study is done on a global scale, and the results should be considered on this scale.

The factors related to the sustainability of urban houses are gathered from the literature and depicted in a long list (see the appendix), together with the number of references mentioning them (i.e., Investigation Frequency (IF)). The complete list of factors with all the references mentioning each can be found, as a supplementary

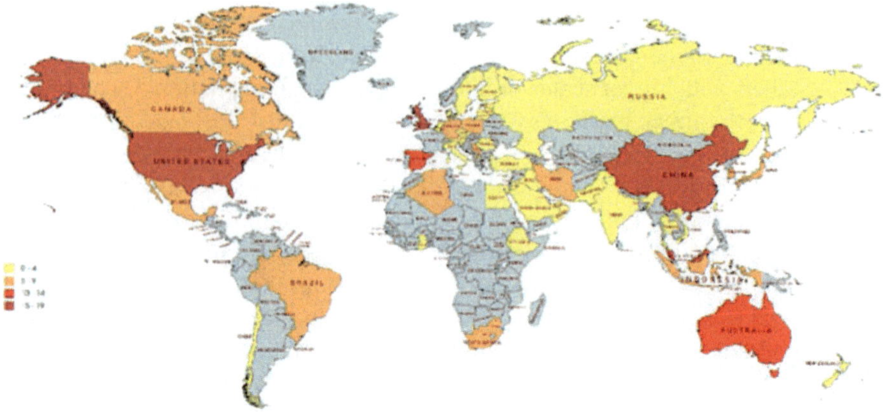

Fig. 2 Distribution of the literature based on the addresses of the authors

file, in [25]. They are prioritized based on IF and Investigation Coefficient (IC). The factors with frequency numbers less than three are combined as the "other" category, which is excluded from analyses.

The numerical values for factors' IFs and ICs are shown in Table 1. Table 1 also shows each factor's scale(s) and area(s) of influence. The overall mean values for IF, and IC (MIF and MIC) for each influence scale (building, neighborhood/community) is illustrated in Table 2. MIF and MIC values are also shown in Table 3 for sustainability sectors that are affected directly by the factors. It should be noted that the factors which impact one sustainability dimension may also affect other sections. For example, the factors considered for the environmental section may or may not affect other aspects. MICs and MIFs are not dependent on the number of factors (from the total of 98 factors) that are related to each scale/sector of influence and, therefore, are calculated as values showing the degree of importance that the studied papers have given to different scales/sectors of sustainability.

As can be seen from the previous table (Table 1), a total number of 98 factors are extracted from the literature and prioritized. According to the results, the top five investigated factors in the literature are "natural resource or energy consumption/efficiency of the building/equipment (during the construction, operation, etc.)," "materials performance (durability, cost, thermal capacity, permeability, ability to re-use, recycled, eco-friendly materials)," "access to public services/infrastructure: availability/quality of services and/or distance/time of travel time to the services (public transport, education/health/shopping/leisure facilities, parks, etc.)," "building spatial layout (size and dimensions, building form, internal space distribution, etc.)," and "location – site – development land" with the investigation frequency numbers of 87, 78, 72, 72, and 67 respectively. Three factors (F1, F2, F5) impact building and neighborhood/community scales, while each F3 and F4 only impose effects on one scale (neighborhood/community and building scales, respectively). It can also be seen that all five factors directly impact environmental and economic sustainability, while F3, F4 and F5 affect the socio-cultural section. None of these

Table 1 The scales/sectors of influence for urban housing sustainability factors and their investigation frequency/coefficient numbers

No.	IF	IC	Building Effect	NBH/Community Effect	Envr. Sust.	Soc-Cult. Sust.	Econ. Sust.	Inst. Sust.
F1	87	0.74						
F2	78	0.66						
F3	72	0.61						
F4	72	0.61						
F5	67	0.57						
F6	66	0.56						
F7	66	0.56						
F8	62	0.53						
F9	59	0.50						
F10	56	0.47						
F11	53	0.45						
F12	52	0.44						
F13	49	0.42						
F14	48	0.41						
F15	48	0.41						
F16	47	0.40						
F17	46	0.39						
F18	43	0.36						
F19	41	0.35						
F20	40	0.34						
F21	40	0.34						
F22	40	0.34						
F23	39	0.33						
F24	37	0.31						
F25	37	0.31						
F26	36	0.31						
F27	35	0.30						
F28	35	0.30						
F29	35	0.30						
F30	35	0.30						
F31	34	0.29						
F32	34	0.29						
F33	34	0.29						
F34	33	0.28						
F35	31	0.26						
F36	31	0.26						
F37	31	0.26						
F38	29	0.25						
F39	28	0.24						
F40	27	0.23						
F41	27	0.23						
F42	26	0.22						
F43	26	0.22						
F44	26	0.22						
F45	25	0.21						
F46	25	0.21						
F47	25	0.21						
F48	24	0.20						
F49	23	0.19						
F50	23	0.19						
F51	22	0.19						
F52	21	0.18						
F53	21	0.18						
F54	20	0.17						
F55	19	0.16						
F56	19	0.16						
F57	18	0.15						
F58	16	0.14						
F59	15	0.13						
F60	15	0.13						
F61	14	0.12						
F62	14	0.12						
F63	14	0.12						
F64	13	0.11						
F65	13	0.11						
F66	13	0.11						

Table 1 (continued)

F67	13	0.11
F68	12	0.10
F69	12	0.10
F70	11	0.09
F71	11	0.09
F72	11	0.09
F73	10	0.08
F74	9	0.08
F75	9	0.08
F76	9	0.08
F77	9	0.08
F78	8	0.07
F79	8	0.07
F80	7	0.06
F81	7	0.06
F82	7	0.06
F83	7	0.06
F84	6	0.05
F85	6	0.05
F86	6	0.05
F87	5	0.04
F88	4	0.03
F89	4	0.03
F90	4	0.03
F91	4	0.03
F92	3	0.03
F93	3	0.03
F94	3	0.03
F95	3	0.03
F96	3	0.03
F97	3	0.03
F98	3	0.03

Grayed cells indicate areas influencing the investigation frequency (IF) and investigation coefficient (IC). Sustainability factors are defined in the Appendix

Note: *IF* Investigation Frequency, *IC* Investigation Coefficient, *NBH* Neighborhood, *Envr.* Environmental, *Soc-Cult.* Socio-Cultural, Econ. Economic, *Inst.* Institutional, *Sust.* Sustainability

Table 2 Overall results for factors' influence scales

Values	Factors with building impacts	Factors with NBH/ community impacts	Factors with impacts on both building and NBH/ community
MIC (Mean Investigation Coefficient)	0.24	0.24	0.29
MIF (Mean Investigation Frequency)	28.65	28.14	33.69

Note: *NBH* Neighborhood

factors has a direct influence on institutional sustainability issues. At the bottom of the list, there are 11 factors (F88–F98) which are the least investigated ones, with IF values of 4 or 3 and IC values of 0.03.

As depicted in Table 2, MICs for the factors with impacts on building and NBH/ community scales are equal (0.24) while higher (0.29) for the factors that impact both scales. This shows that, in the housing sector, the most important factors are the ones that directly affect both the building and its surroundings. Consequently, these factors should be taken into consideration during the design, operation and all

Table 3 Overall results regarding the factors' sectors of influence

Values	Factors with Envr. impacts	Factors with Soc-Cult. impacts	Factors with Econ. impacts	Factors with Inst. impacts	Factors with Envr. + Soc-Cult. + Econ. impacts	Factors with Soc-Cult. + Econ. + Inst. impacts	Factors with impacts on all sectors
MIC	0.28	0.20	0.25	0.19	0.25	0.19	0.21
MIF	33.57	23.22	29.97	22.23	29.57	22.56	24.50

Note: *MIC* Mean Investigation Coefficient, *MIF* Mean Investigation Frequency, *Envr.* Environmental, *Soc-Cult.* Socio-Cultural, *Econ.* Economic, *Inst.* Institutional

the actions regarding both houses and neighborhoods, as they impact both scales. Examples of this could be the development and planning of the local area.

The results also prove that paying attention to the building's surrounding context is as important as the building itself issues to have a sustainable house. This should be applied to all housing-related procedures, such as architectural design and management, housing evaluation methods, and SATs. Although housing assessment tools consider some aspects regarding the neighborhood scale, the importance given is lower than the indicators which affect the building itself. It should be mentioned that applying these factors to the housing assessment processes faces many challenges, such as increasing the cost and time of the assessment. Therefore, there should be solutions to decrease these negative aspects.

According to Table 3, the highest MIC value is allocated to factors with environmental impacts (0.28). Also, the lowest value goes to the indicators with institutional effects (0.19) and those that affect socio-cultural, economic, and institutional sustainability sectors together (0.19). The MIC value for measures with impacts on all sustainability sections is 0.21, and this rate for measures that affect environmental, socio-cultural, and economic sectors is higher and is equal to 0.25. It should be noticed that in highlighting and categorizing the factors based on their influence scale(s)/sector(s), the indicators must affect the targeted area, while they may, as well, have other impacts on another scale (s)/sector(s).

As expected, of the four sustainability sectors studied, the environmental section is the most important one, according to the recent literature. A significant amount of this importance level can be related to energy matters that are studied a lot in the literature and discovered to a great extent in practice. Table 3 shows that, after the environmental factors, economic issues have the most significance in the sustainability of houses compared to the two remaining sectors. Socio-cultural and institutional factors come next, where the least important isthe institutional sector. Although it is expected that the measures affecting all sectors should be of great importance, their importance level is almost 27% less than environmental issues and 18% less than economic factors. However, the factors influencing the three aspects of environmental, socio-cultural, and economic sustainability are the most important factors after the environmental ones, with only 1% of less significance (regarding their MIFs). The reason for that could be the closeness of this combination of sectors with the combination of the well-known pillars of sustainability (environmental, social, and economic pillars). In summary, although many authors have tried to implement holistic approaches in recent years in their works, it is pointed out that different sustainability sections are not given similar degrees of importance in the literature worldwide. This difference between the actual approach of the recent literature and the theoretical strategy of providing the same amount of significance to all sustainability dimensions is considerable.

4 Conclusion

Housing is no longer recognized as a simple shelter for us. It has invaluable importance in our quality of life and well-being. Also, it plays an essential role in the sustainable development of the communities. Therefore, comprehensively knowing what factors contribute to housing sustainability is crucial. This project tries to provide required and valuable information that can be used for two purposes: first, the sustainability evaluation of houses, and second the evaluation of existing housing-related sustainability assessment methods and SATs. The present study has gathered and categorized 98 factors from 118 selected papers from recent years. As the studied literature is from all over the world, the results are global. The listed factors are mostly categories of some housing sustainability indicators and measures, so they may not be used to measure buildings' actual performance. However, these factors and their prioritization can be critical in decision-making-related procedures to act in line with the actual state of knowledge in this area.

We have also investigated each factor's influence on the scale (building and neighborhood/community) and sector (environmental, socio-cultural, economic, and institutional dimensions). The overall degrees of importance for scales/sectors are also calculated. These data can help evaluate SAT indicators and the measures used by other assessment procedures. These results, as well, can lead the related policy- and decision-making processes in a better orientation lined with the current state of knowledge. According to the results, the top five investigated factors in the literature are "natural resource or energy consumption/efficiency of the building/equipment (during the construction, operation, etc.)," "materials performance (durability, cost, thermal capacity, permeability, ability to re-use, recycled, eco-friendly materials)," "access to public services/infrastructure: availability/quality of services and/or distance/time of travel time to the services (public transport, education/health/shopping/leisure facilities, parks, etc.)," "building spatial layout (size and dimensions, building form, internal space distribution, etc.)," and "location – site – development land" with the investigation frequency numbers of 87, 78, 72, 72, and 67, respectively.

Based on the results, environmental aspects get the most importance (more than the factors influencing all sustainability sectors), whereas institutional factors receive the least attention. Also, the factors with impacts on both building and neighborhood scales are the most significant ones. Another interesting result is that neighborhood/community-related issues are considered almost equally important as the building scale factors in the overall sustainability assessment of houses. Therefore, it is necessary to line up the housing- and neighborhood-related policies and strategies and SATs as well, with the importance of neighborhood/community factors dealing with housing sustainability. Also, the analysis outcomes show an opposition between the approach used in the recent housing-related literature on a global scale and the theoretical suggestions regarding giving a similar degree of importance to different sustainability dimensions.

The research has some limitations regarding its comprehensiveness (e.g., not directly including the literature published before 2015 or the ones not indexed in Scopus, not considering the literature written in languages other than English). Decreasing these limitations in future studies can be of use to validate the results. To have a more inclusive view of the factors affecting the sustainability of housing, it could be helpful to analyze the importance of entirely behavioral factors (such as resource consumption behaviors) in such a broad way and improve the validity of the presented results by combining the analyses outcomes. The second suggestion is to widen the scope of the investigation to include non-recent studies. Also, prioritizing the results of this study in different local contexts can be useful to understand better the role of local circumstances in the significance of sustainability factors.

Acknowledgment This research is done as a part of the first author's Ph.D. project, funded by the Mexican National Council for Science and Technology (CONACYT).

Appendix

Prioritized list of urban housing sustainability factors based on the number of references mentioning them (IF).

No.	Factors	IF
F1	Natural resource or energy consumption/efficiency of the building/equipment (during the construction, operation, etc.)	87
F2	Materials performance (durability, cost, thermal capacity, permeability, ability to re-use, recycled, eco-friendly materials)	78
F3	Access to public services/infrastructure: Availability/quality of services and, or distance/time of travel to the services (public transport, education/health/shopping/leisure facilities, parks, etc.)	72
F4	Building spatial layout (size and dimensions, building form, internal space distribution, etc.)	72
F5	Location – Site – Development land	67
F6	Healthy conditions (hygiene, clean environment, air/water quality, mental health, etc.)	66
F7	Housing affordable purchase/rental/mortgage costs (market value, relation to household income)	66
F8	Waste management/facilities (waste recycling/reduction, appropriate waste disposal, etc.)	62

No.	Factors	IF
F9	Safety and security	59
F10	Building equipment/technologies (heating/cooling systems, ventilation systems, kitchen appliances, furniture, etc.)	56
F11	Rehabilitation/refurbishment of the building/community (repairing the deteriorations, functional improvements, etc.)	53
F12	Building envelope (thermal performance of the building, insulation, air tightness/ exchange, etc.)	52
F13	Indoor environment (air quality, humidity, mold, thermal comfort, air circulation, etc.)	49
F14	Carbon footprint – GHG emissions	48
F15	Water management (consumption rate, irrigation systems, recycling, etc.)	48
F16	Use of renewable/clean resources (solar, wind, geothermal, renewable material, etc.)	47
F17	Opportunity for social cohesion/integrity/interaction/connectivity (common use areas, facility sharing, etc.)	46
F18	Re-use/recycle (materials, water, waste, etc.)	43
F19	Housing-related policies/strategies/guidelines/plans/decision-making procedures	41
F20	Building's basic services (safe drinking water availability, access to electricity, sewer, sanitation, etc.)	40
F21	Noise level – Acoustic design – Aural comfort	40
F22	Building codes/energy standards/technical specifications/regulations	40
F23	Investment/finance measures (subsidies, financial risk/support options, investment cost, return of investment, payback period, profitability, cost-benefit data, budget adaptability)	39
F24	Pollution (air, water, land)	37
F25	Climatic/microclimatic conditions (air temperature, humidity, wind speed, solar radiation, heat island effect, etc.)	37
F26	Land use (mixed-use building/community, zoning plans, re-using a developed area instead of new developments, land use change, amount of land supplied, etc.)	36
F27	Natural light – Solar radiation (availability, intensity, etc.)	35
F28	Built-up density	35
F29	Building typology (single-family, attached, apartment, etc.)	35
F30	Flexibility/adaptability (design, construction, function)	35
F31	Neighborhood spatial layout (street layout and network, space between blocks, pedestrian paths, open space layout, human scale features, public furniture, disabled accessibility, etc.)	34
F32	Construction method/techniques/technologies (prefabrication, light/heavy structure, energy efficient techniques, traditional method, etc.)	34
F33	Passive/green/low-energy/near-zero-energy/plus energy design/principles	34
F34	Lighting systems (indoor, street, open space, etc.)	33
F35	Housing occupancy rate – Community population	31
F36	Overall/lifecycle costs of the building	31
F37	Operation cost (energy/water/telephone bills, technology investment price, etc.)	31
F38	Aesthetical quality	29
F39	Traffic – Car dependency – Parking area	28
F40	Space functionality	27

No.	Factors	IF
F41	Participatory actions (design, management, bottom-up governance, educational programs, etc.)	27
F42	Housing/community administration and management (cost/time/risk/maintenance management, etc.), and the types (self-managed, co-managing, etc.)	26
F43	Accessible house (easy physical accessibility for pedestrians/cars/elders/disabled/ etc.)	26
F44	Structural quality and durability	26
F45	Natural hazards and the related resilience/repair (earthquake, flooding, etc.)	25
F46	Building orientation	25
F47	Natural ventilation	25
F48	Walkability/bikeability (auto-free zones, sidewalks, bike routes, etc.)	24
F49	Access to workplaces (distance/time of travel)	23
F50	Local materials	23
F51	Construction cost (material, transport, labor, equipment and installation, etc.)	22
F52	Household/project team overall satisfaction rate	21
F53	Employment/business activity rate/opportunities in the area	21
F54	Compatibility with household/community cultural values or heritage	20
F55	Shading options – Rain protection	19
F56	Building's/neighborhood's identity/reputation/popularity	19
F57	Private/semi-public outdoor space (courtyard, garden, greenhouse, green roof, etc.)	18
F58	Privacy	16
F59	Presence/preservation of cultural heritage/natural resources (ponds, preserved greenery, topographical contours, etc.)	15
F60	Pleasant view/scenery	15
F61	Maintenance cost	14
F62	Renovation/repair/reconstruction cost (material, transportation, etc.)	14
F63	Greening the building (plants, green wall/roof, garden, etc.) and types of greening/ plants	14
F64	Light pollution/quality – Visual comfort	13
F65	Innovation (design, management, technologies, etc.)	13
F66	Socio-cultural mixing of the community	13
F67	Fire prevention/emergency measures	13
F68	Building age – Year of construction	12
F69	Cost/value of land – Land use rights	12
F70	Household transport costs	11
F71	Type of tenure (private ownership, shared/private rent, etc.)	11
F72	Property value retention – Balanced housing market – Market trends	11
F73	Life expectancy of housing – Long-lasting house	10
F74	Biodiversity/wildlife in the area	9
F75	Cater for senior citizens/disabled	9
F76	Ease of movement inside the building (elevators, stairs, furniture and decoration placement, etc.)	9
F77	Construction time/speed	9
F78	Smart home/community (smart technologies/equipment: Energy management systems, smart communication, intelligent controlling of home performance, smart toilets, etc.)	8

No.	Factors	IF
F79	Sense of belonging	8
F80	Odors – Olfactory comfort	7
F81	Access to the city center/urban space (distance/time of travel)	7
F82	Skilled/local labor and, or manager	7
F83	Security of tenure	7
F84	Standards of living	6
F85	Administration/government/management/design cost	6
F86	Ease of maintenance/cleaning (space, equipment)	6
F87	Level of physical deterioration	5
F88	Green/electric car usage – Carpooling	4
F89	Community acceptance/opposition to the project	4
F90	Economic mixing of the community	4
F91	Demographic/ethnic mixing of the community	4
F92	Diversity of building typology and, or spatial/aesthetic forms in the area	3
F93	Openness/closeness of the community (open/semi-open/gated neighborhoods)	3
F94	Community agriculture/gardening	3
F95	Mixed tenure community	3
F96	Access to the internet (speed, capacity)	3
F97	Access to telecommunication service	3
F98	Access to television/cable system	3
	Other	23

Note: The complete list of factors with all the references mentioning each can be found in [25]

References

1. Ibem, A., Aduwo, B.E.: A framework for understanding sustainable housing for policy development and practical actions. In: Architects Colloquium. Architects Registration Council of Nigeria (ARCON), Nigeria (2015).
2. UN-Habitat: Sustainable housing for sustainable cities: A policy framework for developing countries. UNON, Nairobi (2012).
3. Chiu, R.L.H.: Socio-cultural sustainability of housing: a conceptual exploration. Housing, Theory and Society 21(2), 65–76 (2004).
4. Cole, R.J.: Building environmental assessment methods: redefining intentions and roles. Building Research & Information 33(5), 455–467 (2005).
5. Yang, J., Yang, Z.: Critical factors affecting the implementation of sustainable housing in Australia. Journal of Housing and the Built Environment 30(2), 275–292 (2015).
6. Church, C.: Sustainable Development Goals and the housing sector: A change is coming, https://www.sustainablehomes.co.uk/sustainable-development-goals-and-the-housing-sector-a-change-is-coming/, last accessed 2019/9/2.
7. Khodadad, M., Sanei, M.: Suggested Solutions to Improve Security in Urban Environment with an Approach of Sustainable Development. In: 4th International Congress on Civil Engineering, Architecture and Urban Development. Shahid Beheshti University, Tehran (2016).
8. Sanei, M., Khodadad, S., Khodadad, M.: Flexible Urban Public Spaces and their Designing Principles. Journal of Civil Engineering and Urbanism 8(4), 39–43 (2018).
9. Klopp, J.M., Petretta, D.L.: The urban sustainable development goal: Indicators, complexity and the politics of measuring cities. Cities 63, 92–97 (2017).

10. UN-Habitat: Housing at the centre of the New Urban Agenda. United Nations Human Settlements Programme, Nairobi (2015).
11. UN News Center: Affordable housing key for development and social equality, UN says on World Habitat Day, https://www.un.org/sustainabledevelopment/blog/2017/10/affordable-housing-key-for-development-and-social-equality-un-says-on-world-habitat-day/, last accessed 2019/10/6.
12. Ramos, T., Pires, S.M.: Sustainability Assessment: The Role of Indicators. In: Editor, Caeiro, S., Leal Filho, W., et al. (eds.) Sustainability Assessment Tools in Higher Education Institutions, pp. 81–99. Springer, Cham (2013).
13. Mazon, G., Berchin, I.I., Soares, T.C., de Andrade Guerra, J.B.S.O.: Importance of Sustainability Indicators. In: Editor, Leal Filho, W. (eds.) Encyclopedia of Sustainability in Higher Education, pp. 1–8. Springer, Cham (2018).
14. Mateus, R., Bragança, L.: Sustainability assessment and rating of buildings: Developing the methodology SBToolPT–H. Building and Environment 46(10), 1962–1971 (2011).
15. Tomšič, M., Šijanec Zavrl, M.: Development of a sustainability assessment method for buildings – the OPEN HOUSE case. Facilities 36(1–2), 76–102 (2018).
16. Tupenaite, L., et al.: Ranking of sustainability indicators for assessment of the new housing development projects: Case of the Baltic States. Resources 6(4) (2017).
17. Winston, N., Pareja-Eastaway, M.: On indicators of sustainable housing in the European urban contexts. In: ENHR Conference. European Network for Housing Research, Ljubjana (2006).
18. U.S. Green Building Council.: Leed Rating System, https://www.usgbc.org/leed, last accessed 2020/10/2.
19. Sanei, M., Khodadad, M., Calonge Reillo, F.: Analyzing LEED scoring system based on the priorities in urban housing sustainability literature. Urbanism Architecture Constructions 13(2), 151–164 (2022).
20. Building Research Establishment Ltd.: What Is Breeam? https://www.breeam.com/, last accessed 2020/10/2.
21. Sanei, M.: The Comparative Analysis of the Scoring System Used in BREEAM International New Construction 2016 and the Recent Trends in Housing Sustainability-Related Literature. Prostor 30(1), 106–25 (2022).
22. Khodadad, M., Moosavi Nadoshan, S.M., Khodadad, S., Sanei, M.: Operative Guidelines for Sustainable Designing of Child-Oriented Architectural Spaces. Journal of Civil Engineering and Urbanism 8(1), 6–11 (2018).
23. Sanei, M., Khodadad, M., Panahi Ghadim, F.: Effective Instructions in Design Process of Urban Public Spaces to Promote Sustainable Development. World Journal of Engineering and Technology 5(2), 241–53 (2017).
24. Spangenberg, J.H., Lorek, S.: The changing contribution of unpaid work to the total standard of living in sustainable development scenarios. International Journal of Sustainable Development 5(4), 461–75 (2002).
25. Sanei, M., Khodadad, M., Calonge Reillo, F.: The Most Influential Factors on Urban Housing Sustainability Based on a Comprehensive Review of the Recent Literature. Figshare. [Dataset]. https://doi.org/10.6084/m9.figshare.20477112.v2

Academic Study of Tourists' Pro-environmental Behaviour: A Literature Review

Xiao Yue and Yee Choy-Leong

Abstract

While research on pro-environmental behaviour has garnered considerable attention, little attention has been paid to tourists' pro-environmental behaviours (TPEB). The concept, dimensional division, and determinants of tourists' pro-environmental behaviours are discussed in this paper. The study discovered that (1) TPEB research requires additional investigation and that its measurement scale does not adequately account for the unique characteristics of the localised context, (2) the majority of literature overlooks the impact of social interaction on TPEB, (3) additional contextual factors affecting pro-environmental behaviour should be considered, (4) a variety of research methods is urgently needed to study TPEB. Several theoretical viewpoints are necessary to study TPEB development. Future research should use qualitative methodologies to focus on TPEB's structural dimensions.

Keywords

Environmental protection · Pro-environmental behaviour · Tourism · Tourists

---

X. Yue (✉)
School of Business and Economics, Universiti Putra Malaysia, Serdang, Selangor, Malaysia

School of Tourism, Guilin Tourism University, Guilin City, Guangxi, China

Y. Choy-Leong
School of Business and Economics, Universiti Putra Malaysia, Serdang, Selangor, Malaysia

J. T. Anderson (ed.), *Advances in Environmental Sustainability*, Springer Proceedings in Earth and Environmental Sciences,
https://doi.org/10.1007/978-3-031-26365-1_9

9.1 Introduction

Tourism is one of the world's most significant economic industries. The sustainable development of tourism has been one of the most critical challenges influencing tourism in recent years. With the advent of mass tourism, the influence of tourists on tourist sites' environment has been more apparent. Tourists' pro-environmental behaviour (TPEB) refers to the actions of tourists who actively work to mitigate adverse environmental impacts or promote behaviour sustainability throughout the tourism process [1]. The amount of visitors' pro-environmental behaviour is strongly tied to the environmental quality of tourist destinations, which affects the destinations' sustainable development. Many tourism scholars have devoted considerable attention to pro-environmental behaviour, which has developed into a significant research hotspot and frontier in worldwide tourism research.

9.2 Discussion

9.2.1 Definition of TPEB

Scholars employ a variety of terms to describe how people and organisations maintain the natural environment in the literature. The extant literature is perplexing in terms of terminology, and no unified terminology widely recognised and used by researchers has been established [2]. As described in Table 1, the concepts and terminology about environmental protection behaviour are closely connected, but the scientific viewpoint and theoretical perspective behind the terms are somewhat different. This article will primarily use the most frequently used words in the existing literature, pro-environmental behaviour (PEB), to describe the behaviour of individuals actively involved in maintaining and protecting the environment.

9.2.2 Dimensional Division and Measurement

Researchers operationalised PEB from various angles to study in-depth examinations of the manifestations and influencing elements of individual PEBs. They produced a series of measurement scales as practical research tools. The dimensional measures of PEB in the available empirical research literature can be classified as single or multiple dimensions. The research perspective may be classified as general, ecological balancing, social sports, or leisure tourism, as indicated in Table 2.

Most scholars measure TPEB using a single dimension, which does not fully account for or forecast its fundamental qualities. As a result, more research on the multidimensional measurement of TPEB is necessary. Second, numerous scholars assess TPEB using visitors' self-reports. While self-reporting may provide a more accurate picture of visitors' TPEB performance, there is a possibility that tourists purposefully overestimate their TPEB performance. Thus, future research may use self-assessment and other evaluation approaches to improve the TPEB scale

Table 1 Relevant terminologies and definitions of tourist's pro-environmental behaviour

Terminology	Sources	Reference
Pro-environmental Behaviour	An individual makes a conscious effort to lessen the negative influence of their actions on the natural and built ecosystems.	[3]
	A person's or group's behaviour promotes sustainable resource use.	[4]
	Individual or group action is performed to reduce adverse environmental effects.	[5]
	People visit a site less frequently or never when it needs environmental restoration.	[6]
Environmentally responsible behaviour	ERB is any action, individual or communal, aiming to address environmental issues/problems.	[7]
	Be defined as an individual or group action promoting sustainable or decreased resource usage.	[8]
	Individuals or groups that care about the environment try to prevent or remedy environmental issues.	[9]
	Environmentally friendly individual or group behaviour.	[10]
Environmentally friendly behaviour	A tourist's purchasing of eco-friendly tourism goods and services	[11]
	Tourists undertake activities to reduce negative environmental impacts and promote environmental protection in line with particular destinations or scenic sites.	[12]
Environmentally concerned behaviour	Making a difference in the natural environment.	[13]
Environmentally significant behaviour	Environmentally appropriate behaviour is defined by its consequences: how it affects material or energy availability in the environment, ecosystem structure and dynamics, or the biosphere itself.	[14]
Ecological/ecological civilisation behaviour	Individuals attempt to reduce adverse natural effects and contribute to the environmental environment.	[15]
	A set of initiatives encourages tourists to consume fewer natural resources while travelling and relaxing and improve resource sustainability.	[16]

measurement [25]. Numerous researchers' TPEB scales borrow from pre-existing foreign scales [26] without fully integrating the foreign scales into the localised context. To conduct an efficient examination of visitors' TPEB characteristics, it is critical to refine the primary components via qualitative research.

9.2.3 The Determinants of Tourists' Pro-environmental Behaviours

Existing PEB research examines the factors of pro-environmental behaviour and intention. Studying PEB driving factors provides the information base for developing suitable environmental protection management plans and management study

Table 2 Summarising results concerning dimensional measurements of PEB

Dimensionality	Perspective	Dimension division	Reference
Uni-dimensional	General	Actions in education, finance, law, health, and persuasion	[17]
	General	Consumer behaviour, sacrifice intention, and environmental citizenship	[14]
	General	Physical, economical, persuasive, civic, pro-environmental, sustainable, environmentally friendly behaviour	[18]
	General	Practice, convince, and protect	[19]
	Ecological	Social virtues, ecological behaviour, water and energy conservation, eco-conscious consumer behaviour, waste reduction, volunteering for environmental causes, and eco-friendly driving	[20]
		Consciousness of the environment, conservation and social behaviours	[21]
		Low-effort versus high-effort environmental behaviour	[4]
Multi-dimensional	General	General versus specific environmental behaviours	[9]
	Ecological	Intention to conserve and promote the environment	[12]
		Environmental behaviour and compliance tendencies	[22]
		Ecosystem management, green consumerism, intention to pay, and behaviour	[23]
		Positive environmental behaviours and actions, general environmental behaviours and actions	[16]
	Tourism	Sustainable, pro-environment, eco-friendly behaviour	[18]
		New ecological paradigm, consequences awareness, responsibility personal norm, valence, instrumentality, expectancy	[24]
	Social	Consumer behaviour, sacrifice intention, and environmental citizenship	[14]

evaluations, making it a long-standing research relationship [27]. Several exploratory and empirical studies have found factors that promote or influence people's pro-environmental behaviours (Table 3). This section summarises the existing research on antecedent variables from the viewpoints of personal and environmental factors, as well as the theoretical views and assumptions driving the relevant research.

Early literature seldom studies the impact of environmental emotions on environmental behaviours but rather the impact of environmental attitudes or eco-attitudes. Most research has concentrated on the direct influence relationship between local attachment and tourist willingness, with little discussion of the theoretical logic suggested in the relationship [2]. So future research should examine emotions and habits further and try to incorporate them into the existing theoretical framework to explain better the internal process driven by individual environmental responsibility behaviour [27]. Besides, no systematic study of contextual factors

Table 3 Prior research on the effect of causal variables on pro-environmental behaviour

Factor type	Author and time	Method	Causal variables	Guiding theory	Research question(s) and key findings
Personal factors	Bamberg and Möser (2007) [29]	Archival datasets	Archival datasets Problem awareness, internal attribution, social norm, feelings	Norm-activation model, Theory of Planned Behaviour	Personal moral norm is the third predictor of PEB intentions, after attitude and behaviour control.
	Wang (2015) [26]	Survey	Emotion, motivation	Theory of inter-personal behaviour	Personal Norm is an essential factor influencing travellers' intention to travel in civilisation.
	Noppers et al. (2014) [30]	Survey	Instrumental, symbolic and environmental attributes	Adoption of sustainable innovation	The "two-factor theory of emotion and behaviour."
Contextual factors	Sophia, Khorshed, Narelle (2014) [28]	Survey	Tourism enterprises; protected area authorities; local community members; and tourists. egoistic	Stakeholder theory	Environmental attitude influences the degree, direction, and persistence of desire to reduce carbon emission behaviour.
	Vining and Ebreo (1992) [31]	Longitudinal survey	Voluntary curbside recycling program	Norm-activation model	Adoption of sustainable technologies depends on instrumental and symbolic features, including interest in, acceptability of, and intention to adopt.
Contextual and personal factors	Cho et al. (2013) [32]	Survey	Perceived cultural environment (collectivism individualism)	Value-Belief-Norm theory, cultural dimensions	Tourists, followed by tourism businesses, showed a high degree of interest in selfish values.

exists [27]. The study of tourist destination contextual factors is limited [28]. As a result, future researchers should investigate additional antecedent variables, integrate numerous theoretical views, and experimentally evaluate the TPEB creation mechanism integration model [25].

9.3 Conclusion

TERB research needs further research, and its measurement scale does not fully represent the local context. Much research has been conducted on environmental responsibility but less on visitor environmental responsibility [25]. In general, there is a lack of research on environmentally conscious tourism behaviours [2]. Because the majority of pro-environmental research in Eastern countries is based in a Western social context, many scales are clearly out of step with national conditions and social-cultural norms in Eastern countries, and relevant research conclusions may lack the same explanatory power in some contexts [2]. On the other hand, many scholars in eastern countries directly employ the scales and findings of foreign scholars [22], which frequently results in inaccurate experimental results and research conclusions [25]. Thus, future research should focus on elucidating the structural dimensions of TPEB using qualitative methods (such as basic theoretical methodology) and constructing a reliable and accurate TPEB scale through quantitative procedures.

Secondly, most studies ignore the impact of social interaction, namely tourist participation, on pro-environmental behaviour. Additional factors examined in previous research on environmental behaviour include tourist attitudes, perceived behavioural control, and environmental behaviour intentions; results awareness and responsibility attribution; environmental knowledge; and environmental behaviour intentions. The majority of scholars consider that environmental responsibility is a "personal characteristic" of tourists and hence focus only on the internal factors influencing tourists' environmental responsibility [2]. Historically, pro-environmental research has paid little attention to what is perhaps the most basic social medium: social interaction [33]. Individuals seek a sense of belonging and want interaction with others [34]. Individual research, on the other hand, will be studied in terms of "inadequate socialisation" and "oversocialization" [12]. While some scholars are concerned with social ties, others focus on tourists and tourist attractions [35], tourists and residents, and tourists and service providers [36]. Only a few researchers have examined tourism. While completing practices, individual practitioners form bonds with one another [33]. According to some researchers, future research should examine the impact of interpersonal interactions within groups, peer influence, and other factors on visitors' environmental behaviour [37].

Thirdly, studies on the contextual factors influencing pro-environmental behaviour are often ignored. Contextual factors are those that have a direct impact on how people see things. Tourists are influenced by psychological, physical, and environmental factors [38]. The majority of research on visitors' pro-environmental context ignores the impact of contextual factors on personal behaviours. As a result, many scholars propose that future research incorporate additional contextual factors, such as interpersonal relationships (sanitation workers, tour guides, and peer tourists), multimedia technology, advertising, community expectations, environmental protection facilities, and legal institutions [37].

Fourth, TPEB research necessitates the use of a variety of research methods. However, several researchers have studied TPEB production experimentally. Because actual behaviour data differs from self-reported questionnaire data, there is room for improvement in the legitimacy of measurement findings [25]. Additionally, most scholars evaluate TPEB on a single dimension, which does not fully explain or foresee its significance. As a result, more research into the multidimensional measurement of TPEB is necessary.

Furthermore, existing research is based on static cut-off time surveys, and no longitudinal vertical tracking survey of the same tourist group is available. Quantitative research is the most common kind, but qualitative research is uncommon. Existing quantitative research approaches (such as surveys) must be combined with experimental procedures, interviews, qualitative analysis, and other research strategies to ensure the quality of TPEB research results.FundingThis work was supported by the 2021 Guangxi Philosophy and Social Science Planning Research Project (Grant Number: 21CSH003), Guangxi University Young and Middle-aged Teachers' Basic Research Ability Improvement Project (Grant Number: 2021KY0824) and the Guilin Tourism University Scientific Research Project (Grant Number:2021Z07).

References

1. Yu Xiaoting, Wu Xiaogen, Zhang Yuling, et al. (2015).Factors driving environmentally responsible behaviours by tourists: A case study of Taiwan, China. *Tourism Tribune, 30(7),*49-59.
2. Li Qiucheng (2015).*Study on the Driving Factors of Tourists' Willingness to* Environmentally *Responsible Behaviours* (Doctoral dissertation, TesisPhd, Zhejiang University).
3. Kollmuss, A. & Agyeman, J. (2002). Mind the gap: why do people act environmentally and what are the barriers to pro-environmental behaviour? *Environmental Education Research, 8 (3),* 239-260
4. Ramkissoon H, Smith L D G (2013).Testing the dimensionality of place attachment and its relationships with place satisfaction and pro-environmental behaviours. *Tourism Management,* 36,552-566.
5. Miao L, Wei W. (2013).Consumers' pro- environmental behaviour andthe underlying motivations: A comparison between householdand hotel. *International Journal of Hospitality Management,*32,102-112.
6. Walker, G.J., & Chapman, R. (2003). Thinking Like a Park: The Effects of Sense of Place, Perspective-Taking, and Empathy on Pro-Environmental Intentions. Journal of park and recreation administration, 21, 71-86.
7. Borden, R.J., Schettino, A.P. (1979). Determinants of environmentally responsible behaviour. The Journal of Environmental Education, 10 (4), 35-39
8. Sivek, D. J., & Hungerford, H. (1989–1990). Predictors of responsible behavior in members of three Wisconsin conservation organisations. The Journal of Environmental Education, 21(2), 35–40.

9. Cheng T M, Wu H C. (2015). How do environmental knowledge, environmental sensitivity, and place attachment affectenvironmentally responsible behaviour? J. Sustainable Tourism, 23(4), 557-576.
10. Lee T H, Jan F H. (2015). The effects of recreation experience, environmental attitude, and biospheric value on the environmentally responsible behaviour of nature- based tourists. Environmental Management, 56(1), 193-208.
11. Song, H. J, Choongki L, Kang S K, et al. (2012). The effect of environmentally friendly perceptions on festival visitors decision-making process. Tourism Management, 33(6), 1417-1428.
12. Li, Q.C., Zhou, L.Q. (2014). The impact of social capital on tourists' intention to exhibit environment-friendly behaviours. Tourism Tribune, 29(9), 73-82.
13. Axelrod, L.J. & Lehman, D.R. (1993). Responding to environmental concerns: what factors guide individual action? Journal of Environmental Psychology, 13 (2), 149-159
14. Stern, P.C., T. Dietz, T. Abel, G.A. Guagnano, L. Kalof (1999). A value-belief-norm theory of support for social movements: the case of environmentalism. Human Ecology Review, 6 (2), 81–97
15. Kvasova O. (2015). The big five personality traits as antecedents of ecofriendly tourist behaviour. Personality & Individual Differences, 83,111-116.
16. Zhang, S. (2017). What keeps Chinese from recycling? Resour. Conserv. Recycl.,109, 176-186.
17. Smith-Sebasto, N.J., D'Costa, A. (1995). Designing a Likert-type scale to predict environmentally responsible behaviour in undergraduate students: a multistep process. The Journal of Environmental Education, 27 (1), 14-20
18. Lee, T.H., Jan F.H., & Yang, C.C. (2013) Conceptualising and measuring environmentally responsible behaviors from the perspective of community-based tourists. Tourism Management, 36 (3): 454-468.
19. Tonge J, Ryan M M, Moore S A, et al. (2015).The effect of place attachment on pro-environment behavioural intentions of visitorsto coastal natural area tourist destinations. Journal of Travel Research, 33: 356-362.
20. Kaiser, F. G. (1998). A general measure of ecological behaviour. Journal of Applied Social Psychology, 28(5), 395-422.
21. Kaiser, F.G., Wilson, M. (2004). Goal-directed conservation behaviour: the specific composition of a general performance. Personality and Individual Differences, 36, 1531-1544
22. Wan Jicai, et al. (2014). Relationship between specific attributes of place, tourists' place attachment and proenvironment behavioural intentions in Jiuzhaigou. Progress in Geography, 33(3), 411-421.
23. Zeng Feifei, Luo Yanju, Bi Hua, et al. (2014). Ecotourists: Identification and pro-environmental behaviour. Economic Geography, 34(6),182-186; 192.
24. Kiattipoom Kiatkawsin, Han H (2017). Young travelers' intention to behave pro-environmentally: Merging the value-belief-norm theory and theexpectancy theory. Tourism Management, 59,76-88.
25. Qiu H.L., Fan J., Zhao L. (2018). Review and Prospect of Research on Tourists' Environmental Responsibility Behaviour. Journal of Tourism, 33 (11),122-138.
26. Wang Jianming (2015).The Dimensional Structure of Environmental Emotions and Its Impact on Consumption Carbon Emission Reduction Behaviours. Management World, 12,82-95.
27. Steg, L., Vlek, C.(2009).Encouraging pro-environmental behaviour: An integrative review and research agenda. Journal of Environmental Psychology,29,309-317.
28. Sophia Imran, Khorshed Alam, Narelle Beaumont (2014).Environmental orientations and environmental behaviour, Tourism Management, 40, 290-299.
29. Bamberg, S., & Möser, G. (2007). Twenty years after Hines, Hungerford, and Tomera: new meta-analysis of psycho-social determinants of pro-environmental behaviour. Journal of Environmental Psychology, 27(1), 14–25.
30. Noppers, E. H., Keizer, K., Bolderdijk, J. W., & Steg, L. (2014). The adoption of sustainable innovations: Driven by symbolic and environmental motives. Global Environmental Change, 25, 52–62.

31. Vining, J., & Ebreo, A. (1992). Predicting recycling behaviour from global and specific environmental attitudes and changes in recycling opportunities. Journal of Applied Social Psychology, 22(20), 1580–1607.
32. Cho, Y. -N., Thyroff, A., Rapert, M. I., Park, S. -Y., & Lee, H. J. (2013). To be or not to be green: Exploring individualism and collectivism. Journal of Business Research, 66(8), 1052–1059.
33. Halkier, B. (2013). Methodological consequences of integrating concepts of social interaction in practice theoretical approaches. Practice Turn to Praxeological Mainstream Conference, June 6–7.
34. Bernstein, M. J., Zawadzki, M. J., Juth, V., Benfield, J. A., & Smyth, J. M. (2018). Social interactions in daily life. Journal of Social and Personal Relationships, 35(3), 372–394.
35. McGregor, A. (2000). Dynamic texts and tourist gaze: Death, bones and buffalo. Annals of Tourism Research, 27(1), 27–50.
36. Virabhakul, V., & Huang, C. H. (2018). Effects of service experience on behavioural intentions: Serial multiple mediation model. Journal of Hospitality Marketing &Management, 27(8), 99–1016.
37. Liu, J.Y., Wu J.L., Snow, Che T.T. (2019). Understanding perceived environment quality in affecting tourists' environmentally responsible behaviours. Tourism Management Perspectives, 31, 236-244.
38. Miller Z.D., Freimund W.A., Powell R.B. (2018). Measuring elabouration and evaluating its influence on behavioural intentions. J. Interpretation Res., 23 (1), 27-44.

Green Building Design as a Mechanism to Strengthen Resilient and Sustainable Communities: The Case of the Churches and the Stilt Houses, Chiloé Chile

Stefania Pareti, D. Flores, I. Tampe, and L. Rudolph

Abstract

This research aims to explore how strengthening Green Building Design initiatives can facilitate the development of resilient and sustainable communities. The case study selected is the route of the 16 churches declared a UNESCO heritage site and the Pedro Montt axis, which have the main Chiloé stilt houses, since (1) both churches and stilt houses are built in wood, (2) both types of construction are part of the vernacular architectural heritage of the place, (3) they are resilient structures, and (4) the construction of the churches and stilt houses is a crucial element in the development of sustainable and resilient communities.

The methodology starts with the constructive analysis of both cases. It ends with a cadaster of the geographical location of the 16 churches along the island, together with the two main axes of stilt houses in the city of Castro. The results indicate that using native wood as a flagship material adapted to the environment responds to the needs of a humid and rainy climate in a sustainable way. The improvement in the use of wood by the vernacular Chiloé communities allowed the development of complex techniques such as the herringbone box or stake pillars. We conclude that the development of Green Building Design facilitates

S. Pareti (✉)
Universidad Andres Bello, Santiago, Chile
e-mail: s.pareti@uandresbello.edu

D. Flores
Pontificia Universidad Católica de Chile, Santiago, Chile

I. Tampe
Pontificia Universidad Católica de Valparaíso, Valparaíso, Chile

L. Rudolph
Universidad Técnica Federico Santa María, Valparaíso, Chile

J. T. Anderson (ed.), *Advances in Environmental Sustainability*, Springer Proceedings in Earth and Environmental Sciences,
https://doi.org/10.1007/978-3-031-26365-1_10

the development of resilient and sustainable communities, as observed in the case of the Chiloé neighborhoods with their churches and stilt houses that make up vernacular architecture.

Keywords

16 Chiloé UNESCO heritage churches · Chilota school of religious architecture · Construction in wood · Green building design · Resilient communities · Stilt houses

1 Introduction

For this research, we focus on the case of wood as a vernacular material and technique used by Chiloé communities since the seventeenth century. In this framework, our main objective is to explore how strengthening Green Building Design initiatives can be a mechanism to facilitate the development of resilient and sustainable communities.

Developing efficient building systems worldwide has been a fundamental pillar in developing integral communities. Today, the advancing climate crisis represents a design challenge that drives us to find environmentally friendly ways forward. According to the UNEP, 33% of greenhouse gas emissions are directly produced by buildings (i.e., the construction industry is responsible for one-third of greenhouse gas emissions). This problem has come to the professional scene to question the most used construction methods, and construction methods before the industrial revolution that have been proven more environmentally friendly have been re-examined.

Chiloé is an island located in the region of Los Lagos in southern Chile. It is surrounded by the Pacific Ocean and is of heritage value for its richness in traditions, culture, and vernacular architecture. It can be appreciated in the 16 heritage churches recognized by UNESCO and their wooden houses on stilts. The island of Chiloé is characterized by local materials, mainly native wood, and other plant fibers, which are present in the construction of high heritage value and correspond to the study's object.

The arrival of the Jesuits in the seventeenth century and later the Franciscans in the eighteenth and nineteenth century was important to generate the current architecture of Chiloé. The Jesuit influence, together with the knowledge and cultural values of the tenants, allowed them to maximize the use of natural resources within the area. The wood, a fundamental material due to which the construction of the Chilota culture has been based on it since pre-Hispanic times, was a conditioning factor both for the work of the chilotes, as well as for their mentality and their social life, as well as their living, tools, and paths [1]. The above is understood for geographic reasons; its location in the southern hemisphere allows the development of an exceptional forest, with vegetation mainly composed of a variety of trees such as Ciruelillo, Avellano, Coigüe, Ciprés, Canelo, Larch, Ulmo, and Mañío, among others.

The first stilt houses appeared in the twentieth century due to the migratory process from the countryside to the city and birth as a response to a city saturated with urban infrastructure. Thus, mainly linked to fishing, part of the population built their houses on the coast, between land and sea. These villas are characterized by simple volumes with roofs to the waters, with a facade on the sea and another on the street. While residual spaces are used as wood for domestic storage. Furthermore, ecclesiastical architecture and evangelization by the Jesuit community led to the construction of churches within the archipelago. Together with the local population's knowledge, they managed to create structures of significant heritage value and vernacular, mainly using local woods, and designing constructive systems such as assemblies and unique unions.

2 Theoretical Framework

Green Building Designs are buildings that, in their design, construction, and operation process, reduce or even eliminate the negative impacts they generate on the environment compared to traditional buildings. The optimal design of an energy-efficient building has the following objectives: to minimize the total cost of construction and to minimize the environmental impact and energy consumption [2]. Reducing environmental impacts and improving the quality of life are the main reasons for green buildings. At the same time, the associated costs and the lack of government support provided for the change in the direction of the construction industry [3]. However, over the past decade, the design of green buildings has surfaced as a priority. In Rio+20, there is a ten-year commitment to make design a catalyst for public health [4]. This type of practice is not only beneficial for the inhabitants, but it is also beneficial for the security of contractors and their workers, resulting in a better performance of their work under poor favorable conditions [5]. In the case of Chiloé, it is essential to mention using wood as a sustainable material. Currently, materials such as wood promise to reduce the environmental impact of construction. How much reduction can occur is a challenging question, and it depends on various factors [6]. However, more advanced methods using wood are being developed, in addition to looking for diverse construction alternatives. The wood in vernacular architecture has promoted initiatives to create Green Building Design [7].

The architecture and construction in Chiloé are an applied case study. It is undoubtedly constituted by the resources available in the area, and wood is the primary material used in the construction. The woods used are selected according to the functions that must comply with their characteristics, providing rigidity to the structure and thermal insulation. The most resistant woods for the structure, such as Ulmo or coigüe, were in their carved or rough state; for the cypress windows of Mañío and Las Guaitecas, they are also suitable for the covering along with Mañío or larch [8]. These woods are used to construct stilts, structures composed of pierced pillars that arise in response to the context and a need. They are mainly located in bodies of calm waters such as lakes, lagoons, or rivers, without embargo. In the case of Chiloé, they are explicitly found in a coastal border condition [9].

The structures are composed of a series of pillars on water, constituting a constructive technique that arises as a response to the context, located mainly in peaceful water bodies such as lakes, lagoons, or rivers, without embargo. In the case of Chiloé, they are explicitly found in maritime and terrestrial border conditions [9]. The protection that generates the first high level against humidity distinguishes this typology, as is evident in the case of the Niyishi Communal House [10], where the stilt structure and the materiality of its skin-type roofs are justified by them. Snow and heat conditions must be resolved through architecture with the materials available in the area. The houses and corridors of Palafitic Villa in Bad Buchau that make up the town's complex [11] contain elements such as wood and thatch, both resources are available in the area and are valuable as a climate solution.

Around de sus islands, you can find Catholic churches built mainly of wood. These reflect the cultural wealth and Chilota tradition, characterized by the link between local and European knowledge and their constructive techniques [12], related to the environment and keeping alive the spiritual importance of communities. The arrival of the Jesuits meant a process of evangelization of the Chilota indigenous population and the foundations to promote the Catholic Church within the territory [13]. The built churches have attributes linked to the development of constructive techniques and technologies that can adapt to the requirements of the land. In addition, the use of available resources and the knowledge of the places on the wood and the carpentry assemblies are important factors in understanding the churches from a constructive technical point of view.

The island of Chiloé is resilient because it has been affected by natural disasters, mainly earthquakes, due to the seismic condition of the country. The oscillations of the sea, which can reach up to seven meters, also impact stability. Thus, the support structure is designed to work independently of the upper frame, which contains the habitable enclosure [14]. The wood for the supporting structure is from Ulmo, characterized by high mechanical resistance. Barcelona, Spain, is another case of a resilient city, not because of natural disasters but because of exposure to large concentrations of people [15]. Like the case of Chiloé, the demand for tourism in Barcelona is high. Chiloé differs from Barcelona, Spain, as the European policy is more advanced to mitigate the effects of tourism and gentrification. Río de Janeiro, Brazil, is another case of a resilient city. One of the problems experienced in the region is the natural disasters linked to climate: rains and extreme droughts. The rains of 2010 were the key to starting a work of resilience, mainly investing in sustainable city initiatives [16].

3 Methodology

Understanding that the study seeks to plant the vernacular architecture of the Chiloé archipelago as mechanisms of sustainable development for local communities, the first approximation is through constructive and descriptive analysis of both selected typologies of the Chilota architectural school to establish them within the logic of green structures. Additionally, we complement our investigation by studying secondary sources that allow the construction of a theoretical basis necessary for the correct relationship of the concepts to work within the text. This paper also considers the review of public access interviews carried out with the architect Edward Rojas, which, given the years of experience associated with working in the area, sets out a discourse regarding the architecture of the place that it establishes as a facilitator of resilient and sustainable communities.

Therefore, in the first place, a historical and geographical contextualization of Chiloé was carried out, understanding that the vernacular architecture taken as an object of study owes much of its current characteristics to the historical evolution and the geographical features of the island. This will allow us to place both constructive analyses within an environmental context that justifies the decisions taken in the design process, both churches and the stilt houses, understanding how both respond with correct habitability as the primary objective.

The theoretical analysis of the elements was used to complement the characterization of those associated with the constructive analysis for the correct design as generators of sustainable development related to green architecture and the consolidation of resilient communities.

Results The development and consolidation of the imagination of Chiloé in the present day, specifically in its vernacular architecture, is justified and conditioned by the primordial elements: its geography and culture. There are three structuring geographic elements for the archipelago. Omitting the insularity condition, these are the sea, land, and border. From an urban perspective, the sea serves as a limit and a container, in addition to a productive space. The lands become the finite available space for living and producing. The borders become the space for interaction between the above as a diffuse limit [17]. These three conditions govern the entire life of the archipelago. Hence, given the inclemency of the sea and difficulty establishing contact with the continent during the first decades of development on the islands, it is necessary to understand the territory to achieve a level of self-sufficiency that allows you to live without problems in Chiloé.

This concept is related to the second element named in principle: culture. In this sense, the archipelago is characterized by having a cultural collection and particularly associated traditions, which are currently considered one of the many riches of the place. This cultural collection is found in the interactions between native villages and Jesuit missionaries who arrived in the area at the end of the sixteenth and the beginning of the seventeenth century [17, 18]. Their arrival created a cultural syncretism that blended religion, ritual, culture, geography, and constructive knowledge. Part of the syncretism implied the understanding of how to inhabit a geographically and climatically hostile territory while promoting a sense of self-sufficiency. The consolidation of this architectural style allowed the inhabitants to overcome it. Wetness, the rains, and the constant cold necessitated the use of wood as the primary material, reflected in the stilt houses and the churches.

The stilt houses of Chiloé are a classic example of the Chilota living space associated with its vernacular style. The latter is characterized by the construction of simple volumes and low height, where the full predominates over the vacuum concerning the number of spans in the building. The classic configuration was around a stove, positioning the kitchen as a central space, which heated the rest of the house through its constant use. The structure is covered with wooden tiles on the walls, popularly made of alerce given its good behavior in front of the water, to take charge of the rains and avoid filtrations in the interior. In addition to having a clear functional aspect, the wood tiles were conceived as an aesthetic element, an example of the local carpenter's mastery. Another necessary operation is separating the living and the soil to relieve the humidity and allow better ventilation of the home [17]. Therefore, the design decisions are based on the search for thermal comfort within the house as a priority, responding to a context characterized by low temperatures and the constant presence of water.

The stilt houses (Fig. 1) share these characteristics but emphasizes the latter, related to the separation of the soil. Therefore, it was built on the sea using pilots buried in the ocean, mainly by families with few resources associated with fishing, who could not access land on firm ground and therefore built on the shore.

One of the branches that characterized the syncretism mentioned was the ritual, firmly rooted in the current daily life of the archipelago. Thus, indigenous myths and legends coexist with beliefs associated with Catholicism, configuring the Church as the main element within any Chilote settlement. These are constituted as gravitating spaces for their villages, functioning as a center of gravity for the

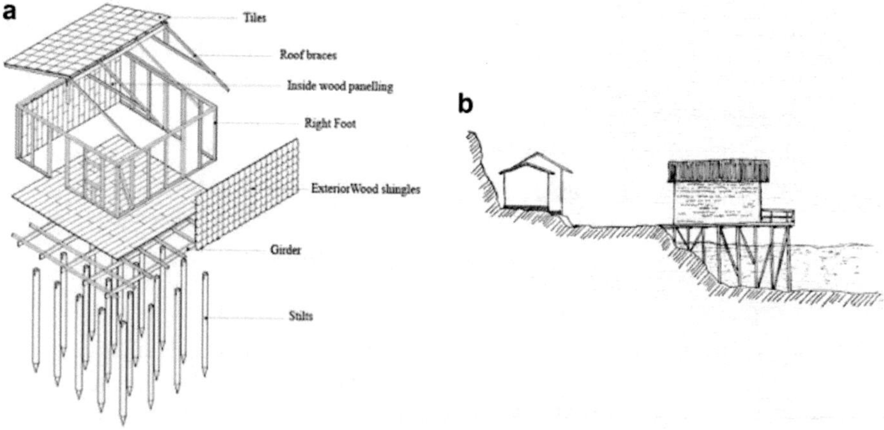

Fig. 1 (**a**) Stilt house constructive analysis. (Own Elaboration). (**b**) Stilt house Sketch. (Own Elaboration)

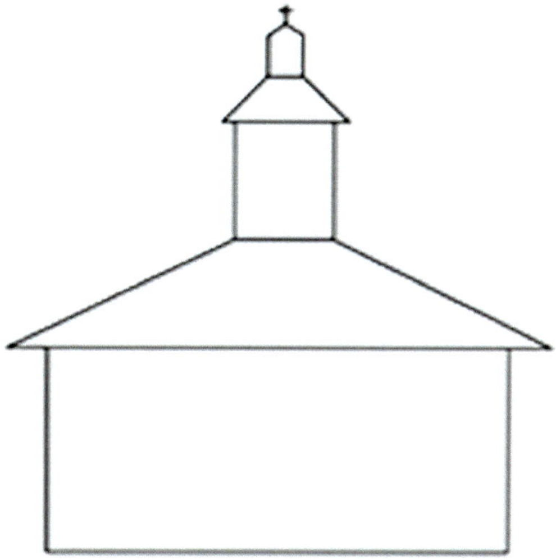

Fig. 2 Basic geometry was used as a typological base for the Chiloé churches. (Own elaboration)

development of each associated locality. In addition, given the relationship with the sea, each becomes a light source, clearly recognizable when approaching some vessel. Thus, the archipelago's life and daily life occur near the las Iglesias church. Approximately 152 churches are within the archipelago, of which 16 were declared Heritage of Humanity by UNESCO [19].

As part of the vernacular, there is a common typological base based on a simple roof geometry and a façade tower to form an access portico (Fig. 2). The interior is configured through three naves (Fig. 3): a central channel with a barrel that ends at the altar and two sides that end at the sacristy and against the vestry [17]. In the same line as the houses, these constructions also take wood as an excellent material, mainly from native species such as cypress, larch, and coihue.

Structurally, they behave as prefabricated elements, working on assemblies, such as box and tenon (Fig. 4), which can be armed and disarmed. Thus, the use of external elements to make unions is relegated to the local master carpenter to fix pieces without needing them (Table 1).

Both objects of study are found within the framework of the consolidation of the Chilota School of architecture, characterized using wood as the primary material [12].

The birth of this style is based on the cultural syncretism of the Chilota culture. Ancestral and local knowledge added to technical knowledge brought from Europe. Knowledge was passed on via carpentry school, with a clear and recognized imprint in the archipelago, reflected in this construction type [17]. The wood on the ground fulfills an aesthetic role and responds to its geographic and climatic context, given its abundance and ability to withstand rain and constant humidity. It is about green structures much before these postulates were widespread.

Added to the material perspective, while the ability to respond to specific needs through local solutions, both constructions become objects of the Chilota culture. These techniques reflected the relationship and understanding of the ecosystem in

Fig. 3 Schematic section of the interior naves. (Own elaboration)

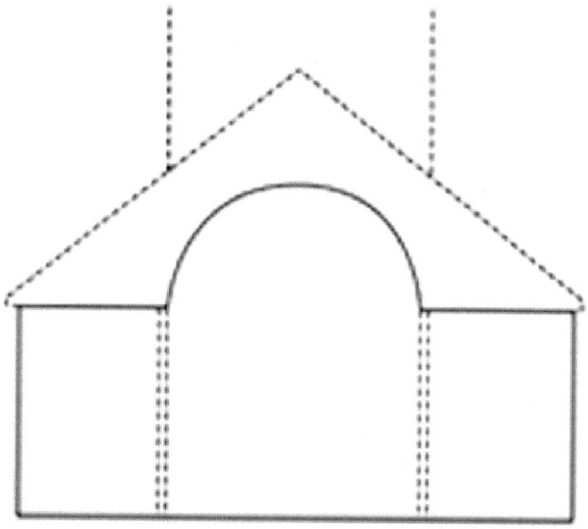

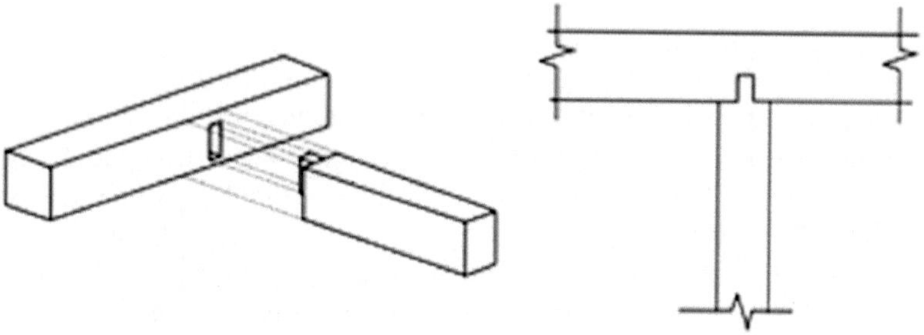

Fig. 4 Box and tenon joint, isometric and section, respectively. (Own elaboration)

which one lives and how to live in it. Architecture becomes an object that builds and reflects identity through understanding and care of the geographic context.

Both concepts are presented as a prominent part of the presented typologies. To adapts them to the current context and take them as models of development for new structures that generate an urban development focused on communities and with a sustainable focus related mainly to low carbon footprint.

An example of this is the programmatic reconversion currently taking place in the Barrio Pedro Montt of the city of Castro, where a large part of the stilt houses in Chiloé are concentrated and where part of these projects are being intervened to host new programs associated with tourism. Thus, the fundamental cultural and constructive characteristics are recognized to establish burial development processes framed in sustainable logics, configuring new constructions that complement ancestral knowledge with new technologies to respond both to the current context and the cultural notions of the element (Table 2).

Table 1 Chiloé school's churches that were selected as world heritage. Own elaboration

Name	Santa Maria de Loreto	Jesus Nazareno	Jesus Nazareno	San Francisco
Location	Achao	Aldachildo	Caguach	Castro

Name	Santa Maria	San Juan Bautista	Nuestra Sefora del Patrocinio	San Antonio
Lo-cati-on	Rilan	San Juan	Tenaun	Vilupulli

Table 2 Chiloé school's churches that were selected as world heritage. Own elaboration

Name	Santiago Apostol	Navidad de Maria	Nuestra senora de la Gracia	Nuestra Seiora de la Gracia
Location	Detif	Ichuac	Nercon	Quinchao

Name	San Antonio	Nuestra Seiora delRosario	Nuestra Seiora delRosario	Nuestra Seiora de los Dolores
Location	Colo	Chelin	Chonchi	Dalcahue

A similar process occurs with the churches. Specifically, the 16 declared Heritage of Humanity, but from a perspective of valuation rather than intervention, thinking about promoting these as part of a sustainable development process in each benefited locality, understanding the value as a tourist object. If this is the case, it is not possible to do so from the intervention, from the restoration and promotion, from the visibility of these within the framework of the routes and their consumption as cultural resources implying benefits for the associated communities, understanding the economic factor that governs tourism and how this can be used for local development.

4 Conclusion

We conclude that the use and improvement of the wood technique by the Chiloé communities shaped an architectural culture characterized by facing new ways of living shaped by a challenging climatic and material environment that pushed them to create new paths in construction. These paths permeated aspects of people's lives, such as housing and religion, hand in hand with the European culture that at one point became intertwined in the Chiloé communities.

This concept opens the way to consolidate critical architectural elements to function sustainably, thanks to wood as a central material. This allows us to point to a greener future focused on sustainability because this material contributes a negative quota of CO_2 by helping in air purification. The rescue of the architectural tradition of ancient cultures promotes a recovery work beyond the merely constructive since the work of others carried out in the past is revalued. The study of these techniques in the company of today's technology can promote new solutions that respond to the problems of the present.

Even so, it should not be forgotten that these processes cannot be carried out only by the associated communities. The presence of these buildings and the efforts for their maintenance over time imply the support of official bodies capable of taking charge of them where the communities fail, assuming their capacities as a limited human group. Thus, the symmetry between actors becomes fundamental for the valuation processes, providing a framework for urbanization that promotes cultural heritage and social development to facilitate the development of resilient and sustainable communities.

References

1. H. Montecinos, Iglesias y Capillas de Chiloé [Churches and Chapels of Chiloe]. Revista de Divulgación Centro Chilote [Chilote center divulgation magazine]. 6, pp. 7–10. (1986)
2. Milajić, A., Milošević, P., & Beljaković, D. (2019). Optimal Green Building Design Using Multi-Objective Genetic Algorithm. Reporting for sustainability, 119–123.
3. Abdelaal, F., & Guo, B. H. W. (2021). Knowledge, attitude and practice of green building design and assessment: New Zealand case. Building and Environment, 201, 107960.

4. Houghton, A., & Castillo-Salgado, C. (2019). Analysis of correlations between neighborhood-level vulnerability to climate change and protective green building design strategies: a spatial and ecological analysis. Building and Environment, 106523.
5. Rajendran, S., Gambatese, J. A., & Behm, M. G. (2009). Impact of Green Building Design and Construction on Worker Safety and Health. Journal of Construction Engineering and Management, 135(10), 1058–1066.
6. Krueger, K., Stoker, A., & Gaustad, G. (2019). "Alternative" materials in the green building and construction sector. Smart and Sustainable Built Environment.
7. Wang, L., Toppinen, A., & Juslin, H. (2014). Use of wood in green building: a study of expert perspectives from the UK. Journal of Cleaner Production, 65, 350–361.
8. Pastor E. El bordemar en la isla de Chiloé. 2018
9. Piesik S. Hábitat: Vernacular Architecture for a Changing Planet. 2017
10. Rojas E, Elmúdesi B, Palafitos of Castro/Chiloé Vernacular contemporary heritage. 2019
11. Maier U, Harwath A, Detecting intra-site patterns with systematic sampling strategies. Archaeobotanical grid sampling of the lakeshore settlement Bad Buchau-Torwiesen II, south-west Germany, 2011.
12. Subsecretaría de Patrimonio Cultural [Undersecretary of Cultural Heritage]. Estudio Inventario de Iglesias del Archipiélago de Chiloé pertenecientes a la Escuela Chilota de arquitectura religiosa en madera. [Inventory Study of Churches of the Chiloé Archipelago belonging to the Chilota School of religious architecture in wood.], Ministerio de las Artes, la Cultura y el Patrimonio [Ministry of Arts, Culture and Heritage], Santiago (2019)
13. R. Urbina, Las misiones franciscanas de Chiloé a fines del siglo XVIII: 1771–1800 [The Franciscan missions of Chiloé at the end of the 18th century: 1771–1800], Elártole Editorial, Valparaíso (1990)
14. E. Pastor, El Bordemar de la Isla de Chiloé. Los palafitos de la ciudad de Castro en relación con el Lugar. 2018
15. ¿Qué son las ciudades resilientes? [Online]. https://www.hola.com/estar-bien/20180524124117/que-son-ciudades-resilientes-gt/ 2018 [Last view: 2021 December, 9th].
16. M. López, Building sustainable communities [Online]. https://www.expoknews.com/que-es-una-ciudad-resiliente-y-8-ejemplos/ 2016 [Last view: 2021 December, 9th].
17. Rojas, E. (2021). Artifica tu casa 2021: Edward Rojas habla sobre la arquitectura de Chiloé [Artífice your house 2021: Edward Rojas talks about Chiloé's architecture].
18. H. Barrientos, s.f Iglesias y Capillas de Chiloé (Revista de Divulgación Centro Chilote, N°6, 7–10 pp.
19. L. Mancilla, "Chiloé entre mares" [En línea]. 2013 https://issuu.com/altolamurta/docs/chiloe_entre_mares [Lat visit: 2021 December, 9th].

Analysis of the *Vicugna pacos* (Alpaca) Wool Fiber in the Properties of Concrete

Levi Acevedo Villazana, Cristian Starky Mora Altez, and Steve Camargo Hinostroza

Abstract

This research aimed to analyze alpaca (*Vicugna pacos*) wool fiber as a reinforcement in the properties of concrete. For this purpose, methodologies such as ASTM (American Society for Testing and Materials), ACI (American Concrete Institute), and Peruvian Technical Standards (NTP) were used; likewise, the experimental design methodology was employed. The sample consisted of 45 specimens; 30 were cylindrical, and 15 were in the form of beams. The alpaca wool fiber was used in percentages of 0%, 0.5%, 1%, 2%, and 4%, with 60 mm length cuts added to the concrete mix. The results showed that the compressive strength in the specimens containing the alpaca wool fiber at 28 days complies with the percentages established by the ACI standard. However, the compressive strength of 33.48% was decreased in the control specimen. As for the tensile strength, compared to the control specimen, an average increase of 1.19% was observed on day 28. Finally, compared to the control specimen, the values obtained in the flexural strength of the alpaca wool fiber specimens showed an average reduction of 19% and 18% on days 14 and 28, respectively. In conclusion, when analyzing the samples containing the alpaca wool fiber, we observed that the tensile strength improved thanks to alpaca wool fibers.

Keywords

Alpaca wool fiber · Compressive strength · Flexural strength · Tensile strength

L. Acevedo Villazana (✉) · C. S. Mora Altez · S. Camargo Hinostroza
Universidad Continental, Huancayo, Peru
e-mail: 44530969@continental.edu.pe; 71910165@continental.edu.pe; scamargo@continental.edu.pe

J. T. Anderson (ed.), *Advances in Environmental Sustainability*, Springer Proceedings in Earth and Environmental Sciences,
https://doi.org/10.1007/978-3-031-26365-1_11

1 Introduction

Worldwide, concrete is considered one of the basic construction materials due to its high strength [1]. In 2021 the concrete industry increased by 10% and is expected to increase by 5% more by the end of 2022 compared to the previous year [2]. However, temperature variation, rainfall, and other pathologies cause the mechanical properties of concrete to decrease significantly [3], and Peru is no exception. According to the National Meteorology and Hydrology Service of Peru, in Junin's Region, during June 2022, the maximum temperature was 32 °C, and the minimum was −4.8 °C. Likewise, the average annual rainfall is 700–2000 mm [4], thus causing the useful life of concrete to be impaired. Because of this, many industries add inorganic and organic fibers to improve their mechanical properties.

The use of natural animal fibers in construction has many advantages due to their high resistance to abrasion, bending, tenacity, compression, and traction, among others [5]. They also mitigate environmental impacts and reduce transportation costs [6]. Natural fibers are tiny filaments found in cotton seeds, flax, wool, hair, or silk; and are considered valuable raw materials for construction [5]. Alpaca (*Vicugna pacos*) fiber is composed of keratin, whose diameter is approximately 18–27 mm [7]. This fiber has the particular characteristic of being more resistant and softer than sheep fiber [8]. Recently, research has been carried out on using natural fibers in concrete, as in Alyousef et al. [9], where their objective was to determine the effects of sheep wool fiber on the properties of concrete. For this purpose, they used wool percentages of 0, 0.5, 1, 1.5, 2, 3, 4, and 6%, which were added to the concrete. The findings showed improved tensile strength (32.7%) and flexural strength (20.8%). They concluded that concrete with wool fiber improves its effectiveness.

Similarly, Alyousef et al. [10] questioned if sheep wool improves concrete transport properties and strength. For this purpose, they made five mixtures using ordinary cement and sheep wool of 70 mm length, ranging from 0–2.5%. This research showed that adding fibers to the concrete decreased its slump. However, it increased the compressive strength by 30 MPa at 90 days after curing. Additionally, they observed that the flexural strength had higher values than the concrete without fiber. The researchers concluded that adding wool fiber improves some properties of concrete.

In their experimental investigation, Ahmad and Rehman [11] used sheep wool to increase the total strength of concrete. They found that the flexural, compressive, and tensile strength increased by 21%, 29%, and 12%, respectively, compared to concrete without adding sheep wool fiber. However, concrete with 5% wool had lower capillary absorption. Al-Kafri et al. [12] determined the effect of sheep wool fiber on the properties of concrete, where the proportion of the concrete mix was 350 kg of cement, 700 kg of sand, 1050 kg of gravel, and 219.8 kg of water. The specimens had percentages of wool fiber at 0%, 0.09%, 0.12%, and 0.14%. They found that the compressive, flexural, and tensile strength improved after 90 days of curing for all the specimens. However, the elasticity decreased as a higher percentage of fiber was added. They concluded that adding the sheep wool fiber improved the properties of the concrete.

Adding natural sheep fibers to the concrete mix benefits the concrete's properties because the fibers are flexible and cover pore spaces. In all the investigations, researchers followed standardized methods and established concrete designs. Regarding alpaca wool fiber, the study conducted by Quispe et al. [13] in South America identified the existence of more than one million small producers who have alpacas, and about 30% of the fiber is used in their communities. About 10% is waste, and these residues are not used in most cases. Therefore, this work proposes to use alpaca wool fiber to determine if it influences the mechanical properties of concrete, hoping that it can contribute to the reduction of waste and considering that concrete mixed with fibers improves the properties of concrete to a great extent [14].

2 Materials and Methods

2.1 Alpaca Wool Fiber Characteristics

Natural fibers, such as alpaca, are used in different industries, and their importance is immense due to their high elasticity and resistance [15]. Peru has the largest alpaca population; it is also the largest producer of alpaca fiber worldwide [16]. By 2017, the alpaca population had increased up to 50.2% compared to previous years [17]. The main properties of alpaca fiber are high fire resistance, high absorption, high water repellency, thermal resistance, insulation, strong thread, odor absorbing, rot resistance, and vibration control [18]. The physical properties of the alpaca fiber from the Huacaya breed are minimally variable (Table 1).

The average diameter of alpaca wool fiber was 31.7 microns (Table 1); likewise, the coefficient of variation of the fiber diameter was 23 microns, which indicates that there is not much difference about the mean. The average fiber length was 94.2 mm. As for the fiber curvature index, on average, it was 26.5 Dg/mm, which indicates that there is no significant difference among the samples.

Table 1 Physical properties of alpaca fiber

	Diameter -Micron[mic]	Standard deviation -SD[mic]	Coefficient of variation -CVD[%]	Comfort factor -CF[%]	Finesse Wool-SF(mic)	Staple Len -SL[mm]	Fiber curvature CRV[Dg/mm]
MWF1	32.9	7.6	23.2	39.1	32.7	95	26.6
MWF2	34.4	7.4	21.5	30.3	33.7	90	24.7
MWF3	33	7.1	21.6	37.1	32.2	100	23.5
MWF4	30.4	7.3	23.9	53.3	30.4	90	27.1
MWF5	29.5	7.2	24.5	59.1	29.5	100	27.2
MWF6	29.7	7.5	25.3	57.9	30.1	90	30.1
Average	31.7	7.4	23.3	46.1	31.4	94.2	26.5

Table 2 Pearson's correlation between the characteristics of alpaca fiber fleece

Pearson's correlation								
		Micron (mic)	SD (mic)	CVD (%)	CF (%)	SF (mic)	Staple Len (mm)	CRV (Dg/mm)
Micron (mic)	Pearson's correlation	1						
SD (mic)	Pearson's correlation	0.702	1					
		0.120						
CVD (%)	Pearson's correlation	0.411	0.277	1				
		0.419	0.595					
CF (%)	Pearson's correlation	0.204	−0.073	0.936[a]	1			
		0.699	0.891	0.006				
SF (mic)	Pearson's correlation	−0.061	0.244	−0.863[b]	−0.985[a]	1		
		0.909	0.642	0.027	0.000			
Staple Len (mm)	Pearson's correlation	−0.612	−0.598	−0.167	0.014	−0.130	1	
		0.196	0.210	0.752	0.979	0.806		
CRV (dg/mm)	Pearson's correlation	0.542	0.510	0.948[a]	0.797	−0.684	−0.416	1
		0.266	0.301	0.004	0.058	0.134	0.412	
	N	6	6	6	6	6	6	6

[a]The correlation is significant at level 0.01 (bilateral)
[b]The correlation is significant at level 0.05 (bilateral)

The Pearson Correlation of the alpaca fiber characteristics had a high negative correlation ($r = -0.612$) between Micron and SL, i.e., the lower the micron (finer the fiber), the higher the SL (Table 2). Similarly, the CF has a high negative correlation with SF, so the SF of low value (higher fineness of the fibers) had a higher CF.

2.2 Mix Design Ratio

The determination of the mix design was carried out by the ACI – 211 standard [19]; where the fineness modulus of the coarse aggregate was 6.70, the nominal maximum size was 3.8 cm, the absorption percentage for coarse and fine aggregate was 2.8% and 5.7% respectively; likewise, the moisture percentage of coarse and fine aggregate were 1.7% and 2.5% respectively. Once the calculations were made, the proportion for the mix design was obtained; these values are shown in Table 3. It is worth mentioning that the cement used complied with ASTM C150–16 [19].

Fig. 1 Concrete preparation with the addition of alpaca wool fiber

Table 3 Mix design for 1m^3 of concrete

Cement	Fine aggregate	Coarse aggregate	Water
350.8 [kg]	629.6 [kg]	1263.1 [kg]	215.3 [L]
Design proportion			
1	1.8	3.6	27 [L]

2.3 Concrete Preparation

Once the design calculations were made, the concrete was prepared to take into account the calculated proportion; these materials were in the mixer as mentioned in ASTM C192-02 [20], then the alpaca wool fibers were added to each beam and cylindrical concrete specimens. For this, the percentage of wool to be added was calculated about the water and cement (0.5) [9]. The concrete beams and cylindrical concrete specimens were reinforced with 0%, 0.5%, 1%, 2%, and 4% with cuts of 6 mm in length; these values were chosen because previous research suggests that the percentages should be less than 6% since, if they are higher, they do not have good results in the mechanical properties [10] (Fig. 1).

2.4 Experimental Test

To determine the compressive and tensile strength of concrete, the ASTM C39/C39M-18 standards [21] and the NTP 339.034:2015 [22] were taken into account for cylindrical concrete specimens, where 30 specimens of 15 cm diameter and 30.5 cm height were prepared for evaluation at 7, 14 and 28 days. Likewise, for the flexural strength tests, ASTMC293/C293M-16 standards [23] and NTP 339.079:2012 (revised in 2017) were taken into account [24] where 15 beams of 61 cm long, 15 cm wide, and high were elaborated (Fig. 2).

Fig. 2 Specimens tests

3 Results

3.1 Compressive Strength Test

At day 28, the compressive strength had values higher than 300 kg/cm^2 without the addition of the alpaca wool fiber; in addition, in the specimens containing the alpaca wool fiber, they managed to maintain the strength with an average value of 205 kg/cm^2, 145 kg/cm^2 and 102 kg/cm^2 at 28, 14 and 7 days respectively. However, the lowest value obtained was on day 7 with 57 kg/cm^2 with a fiber percentage of 0.5%. It should be noted that, on day 28 of curing, an average decrease of 33.48% in compressive strength was observed in the specimens containing the fiber compared to the control specimen. When compared with the percentage of resistance according to the ACI 318 standards [25], it was identified that the specimens containing the alpaca wool fiber at 28 days comply with the established standard (100–120%) (Figs. 3 and 4). On the other hand, the most frequent fractures on day 7 were type III; on day 14, it was type V; on day 28, it was type VI. Therefore, according to ACI standards, adding alpaca wool fiber complies with the compressive strength in the 28 days of evaluation.

3.2 Tensile Strength Test

The control specimen had a resistance factor of 190 kg/cm^2, 235 kg/cm^2, and 294 kg/cm^2 on days 7, 14, and 28, respectively; likewise, the specimen with 4% alpaca wool fiber had higher values regarding the other models containing percentages of wool fiber, where the results were: 208 kg/cm^2, 248 kg/cm^2 and 307 kg/cm^2, for days 7, 14 and 28 days of age respectively. Regarding the percentage of tensile strength compared to the control specimen, an average increase of 1.19% was observed on day 28. Also, it was evidenced that the most common types of fracture in the

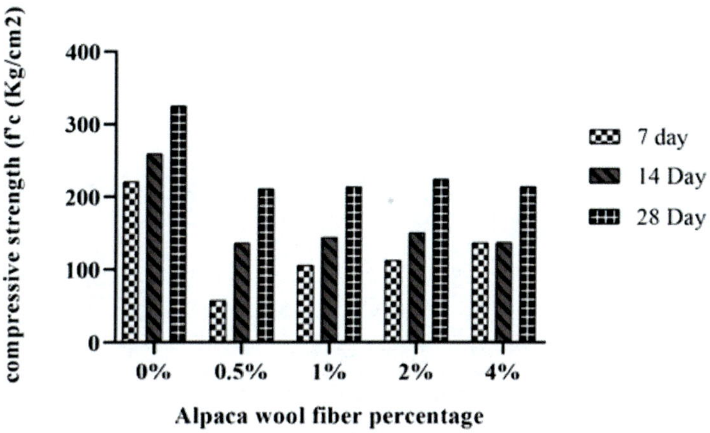

Fig. 3 Compressive strength test on the specimens

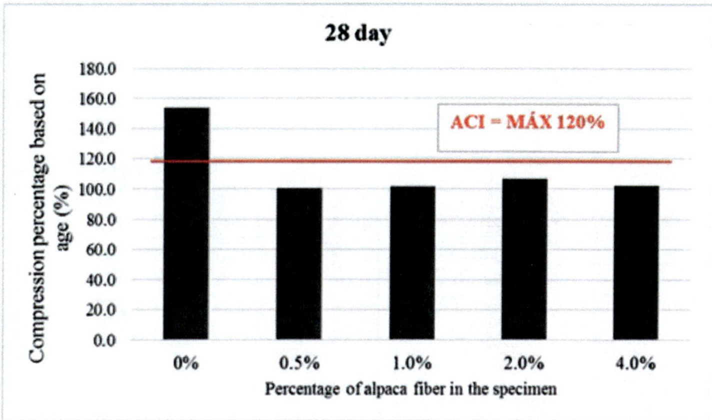

Fig. 4 Compressive strength as a function of age according to ACI standards

samples on days 7, 14, and 28 were type III. On the other hand, when comparing the strength of concrete as a function of age, according to the ACI-318 standards [25], it can be seen that in all the specimens containing the alpaca wool fiber, as well as the control specimen, they exceed the recommendation of this standard, whose maximum suggested value is 120% (Figs. 5 and 6).

3.3 Flexural Strength Test

At 28 days, the flexural strength in the control sample had a maximum value of 51.7 kg/cm^2; at 14 days, the specimen with 0.5% of alpaca wool fiber had its highest value of 41.4 kg/cm^2; at 28 days, the specimen with 1% wool fiber recorded a maximum value of 43 kg/cm^2; the specimen with 2% fiber on day 28 recorded a

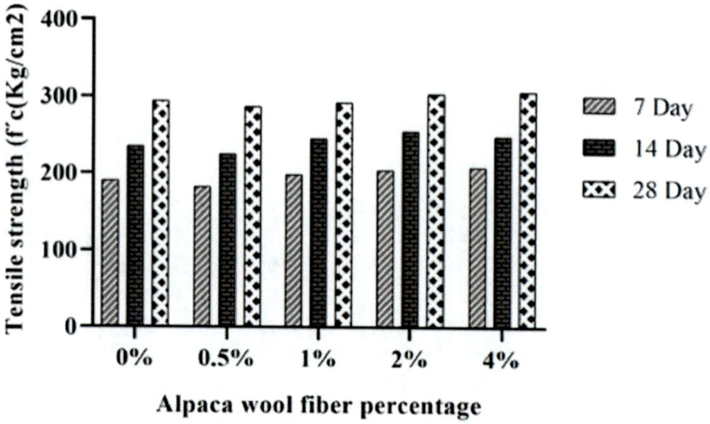

Fig. 5 Tensile strength test of the specimens

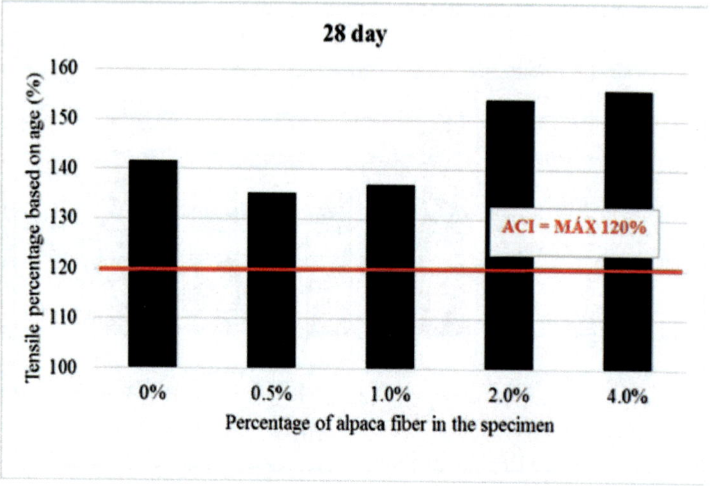

Fig. 6 Tensile strength as a function of age according to ACI standards

maximum value of 45 kg/cm^2; finally, the specimen with 4% fiber had the highest value recorded on day 28 with 42 kg/cm^2. As for the flexural strength compared to the control specimen, there was an average reduction of 19% and 18% on days 14 and 28, respectively. It is worth mentioning that as the percentage of alpaca wool fiber in the concrete increases, the flexural strength decreases since the average variation of the flexural strength was 5 kg/cm^2 (Fig. 7).

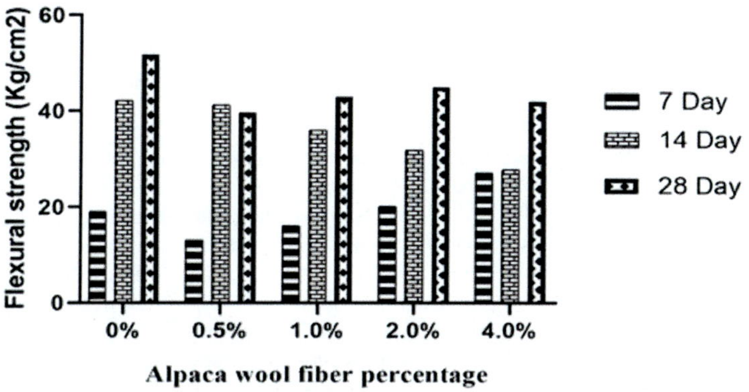

Fig. 7 Flexure strength test of the specimens

4 Discussion

This research showed that the specimens containing the alpaca wool fiber caused the compressive strength to decrease. However, the percentage of compression at the age of 28 days in these specimens complied with the ACI- 318 standards [25]. Regarding the tensile strength, an improvement could be detected at day 28 in all the specimens with alpaca wool fiber; likewise, the flexural had no significant difference in curing at day 28, compared to the control specimen. The findings of this study agree with Alyousef et al. [9] when determining that adding sheep wool fiber improves tensile and flexural strength at 28 days of curing. It also agrees with Gadgihalli et al. [26] when adding sheep wool fiber improved the compressive and flexural strength of concrete;. However, as the cement grade was increased, the strength percentage gradually decreased. Similarly, it agrees with Galema et al. [27] when finding that compressive, flexural, and split tensile strength improved with the addition of sheep wool fiber in the concrete, especially the higher flexural strength with 1.5% fiber. Likewise, Ahmad and Rehman [11] determined that workability was reduced; flexural and tensile strength increased, this thanks to the addition of sheep wool fibers in concrete. Although research shows the importance of sheep wool fiber in concrete reinforcement, not much exists about alpaca wool fiber, mostly simple descriptions of properties and characteristics. This study was able to analyze the attributes of alpaca wool fibers where the high fineness of the thread was evidenced, results that agree with Paucar et al. [17] when identifying the negative correlation between the average diameter of the fiber and the length of the thread; thus determining its high fineness. Likewise, the effects of the alpaca wool fiber on the compressive, tensile and flexural properties of hardened concrete were analyzed, where the thread improved tensile strength and decreased compressively and flexure strengths, on day 28 of curing.

References

1. Bag of concrete. Why is concrete important. https://bagofconcrete.com/why-is-concrete-important/. last accessed 2022/06/03
2. The Economist. The concrete industry grows 10% in 2021 and forecasts a rise of 5% in 2022.https://www.eleconomista.es/empresas-finanzas/noticias/11627315/02/22/La-industria-del-hormigon-crece-un. last accessed 2022/06/03
3. Ghadzali, N., Ibrahim, M., Mohd Sani, M., Jamaludin N, Desa, M., Misri, Z. Properties of concrete containing different type of waste materials as aggregate replacement exposed to elevated temperature – A review. In IOP CONFERENCE SERIES: earth and environmental Science; 2018; Pahang.
4. National Service of Meteorology and Hydrology of Peru. current weather conditions. https://www.senamhi.gob.pe/main.php?dp=junin&p=condiciones-climaticas#. last accessed 2022/06/03
5. Boumaaza, M., Belaadi, A., Bourchak, M., Systematic Review on Reinforcing Mortars with Natural Fibers: Challenges of Environment-Friendly Option. Journal of Natural Fibers 17(4), 12–15. (2022)
6. Kathib, J., Machaka, M., Elkordi, A.: Manual of Sustainable Management of Concrete ansd Industrial Waste New York: Elsevier; 2021.
7. Mcgregor B. Physical, chemical, and tensile properties of cashmere, mohair, alpaca, and other rare animal fibers. ScienceDirect 105–136. (2018)
8. Vaca-Cárdenas, M., Oleas, M., Vaca-Cárdenas, M., Velasco, A.: Characterization of the Physical – Mechanical Properties of Alpaca Fiber (Vicugna pacos) at the Tunshi Experimental Station. The Ecuadorian Journal of S.T.E.A.M. 1(1) 29–56. (2020)
9. Alyousef, H., Alanduljabbar, H., Mohammahosseini, H.: Utilization of sheep wool as potential fibrous materials in the production of concrete composites. Elsevier 2356(7102), 101–216. (2020)
10. Alyousef, H., Mohammasdhosseini, A., Khalek, A.: An Integrated Approach to Using Sheep Wool as a Fibrous Material for Enhancing Strength and Transport Properties of Concrete Composites. Journal Materials. 2022; 15(1688): p. 1–20.
11. Ahmad, I., Rehman, K.: Experimental investigation on using sheep wool as fiber reinforcement in concrete giving increment in overall strength. ScienceDirect 45(6), 4405–4409 (2021)
12. Al-Kafri, A.: Effect of Natural Fiber (Sheep's Wool) on Properties of Fresh and Hardened Concrete. In THIRD INTERNATIONAL CONFERENCE ON TECHNICAL SCIENCES (ICST2020); 2020; Tripoli. p. 28–30.
13. Quispe, E., Rodriguez, T., Iñiguez, R., Mueller, P.: Producción de fibra de alpaca, llama, vicuña y guanaco en Sudamérica. Animal Genetic Resources Information. 2009; 45(2): p. 1–14.
14. Ahmed, M., Hoque, M., Shaid, A.: A Comparative Study of the Mechanical Properties of Jute Fiber and Yarn Reinforced Concrete Composites. Journal of Natural Fibers. 2018; 17(5): p. 676–687.
15. Jankowska, Wyrostek, A., Patkowska, S., Czyż.: Comparison of Physico-mechanical Properties of Fibre and Yarn Made of Alpaca, Sheep, and Goat Wool. Journal of Natural Fibers. 2019; 18(10): p. 1512–1517.
16. National Institute of Statistics and Informatics. National Agricultural Census 2012. Lima:; 2012.
17. Paucar, R., Ruiz, A., Lafraya, S., Mendoza, G., Alvarado, F.: Características textiles de la fibra de alpacas Huacaya (Vicugna pacos). Scientia Agropecuraria. 2019; 10(3): p. 1–14.
18. Agrarian University of La Molina. Directed technical assistance in characterization and classification of alpaca fiber. Huancavelica:; 2012.
19. American Society for Testing and Materials. Standard Specification for portland cement ASTM C150. New York:; 2016.
20. American Society for Testing and Materials. Preparation and Curing of Concrete Test Specimens in the Laboratory ASTM C192-02. New York:; 2018.

21. American Society for Testing and Materials International. Standard Test Method for Compressive Strength of Cylindrical Concrete Specimens1 ASTM C39/C39M-18. New York:; 2018.
22. Technical Standards Peruvian. Concrete. Standard test method for determination of the compressive strength of concrete in cylindrical samples. Lima:; 2015.
23. American Society for Testing and Materials International. Standard Test Method for Flexural Strength of Concrete (Using Simple Beam with Center-Point Loading), ATSM ASTMC293/C293M-16. New York:; 2018.
24. Peruvian technical standard. Concrete. Standard test method for flexural strength of concrete (using simple beam with center-point loading). Lima:; 2017.
25. American Concrete Institute Standar. Regulation Requirements for Structural Concrete (ACI 318S-05) and Comment (ACI 318SR-05). New York:; 2015.
26. Gadgihalli, V., Meena, Y., Chandana, R., Ha, R.: Analysis of properties of concrte using sheep wool dipping in salt water as fibre reinforcement admixture. International journal of research Granthaalayah. 2017; 5(11): p. 1–3.
27. Galena, D., Kebede, G., Feleke, L. Investigation on effects of sheep wool fiber on properties of C-25 concrete. Saudi Journal of Civil Engineering. 2019; 1(1): p. 1–28.

Assessment of Bridge Damage Based on Analytic Hierarchy Process

Han Chen, Mengzhu Liu, and Xiaoning Wang

Abstract

For a comprehensive assessment of the technical condition of bridge structures, combining domestic and international bridge assessment methods, this paper establishes a time-influenced Analytic Hierarchy Process (AHP) to assess bridges in a hierarchical and graded evaluation. At the same time, based on the hierarchy model, we constructed a comprehensive condition assessment system, a judgment matrix for long-span cable-stayed bridges, and the consistency validation method of AHP to determine the reasonableness of the assessment results. We used ANSYS simulation analysis software to validate the method proposed in this paper and used the current structural condition of a cable-stayed bridge as an example. The results show that the current structural condition of the bridge is good, and the assessed structure corresponds to the actual state of the bridge. This paper's bridge condition assessment system is based on bridge wear and time-influenced AHP. This technique is conducive to an accurate grasp of the bridge's operational status and is significant to bridge maintenance and repair.

Keywords

Bridge inspection · Analytic hierarchy process · Time influenced factor · Damage assessment

H. Chen (✉) · M. Liu · X. Wang
College of Transportation Engineering, Dalian Maritime University, Dalian, China
e-mail: hchen@dlmu.edu.cn

J. T. Anderson (ed.), *Advances in Environmental Sustainability*, Springer Proceedings in Earth and Environmental Sciences,
https://doi.org/10.1007/978-3-031-26365-1_12

1 Introduction

Road bridges have made a considerable contribution to the rapid development of the transport industry, and the technical condition of bridges or testing the load-bearing capacity is an essential task for road management and maintenance units. The purpose of bridge inspection is to evaluate the condition of bridges and ultimately to provide a basis and support for the development of bridge maintenance countermeasures. In engineering practice, standard bridge inspections are daily, periodic, or special assessments, with special patrols arranged when necessary. These inspections allow for the timely detection of bridge corruption that occurs during use and the tracking of bridge disorder trends. However, bridge inspection and management systems in China started relatively late, firstly in 1986 when the Ministry of Transport embarked on a study of the China Bridge Management System (CBMS). Still, it was not until 1993 that the system began to be rolled out nationwide [1]. After more than twenty years of continuous progress and improvement, the system now has the functions of recording the condition of bridges, evaluation, analysis, maintenance, and decision-making. However, due to the late start and the lack of profound accumulation of technology and experience, China's bridge management system still has certain shortcomings compared with developed countries. This paper summarizes the classification of underwater damages to bridge structures through a review of data, examines the mechanism of its occurrence, and proposes corresponding assessment criteria for different degrees of disease damage. This paper presents a time-influenced Analytic Hierarchy Process (AHP) to determine the weights of each evaluation index, which improves the efficiency of bridge inspection and reduces work bias.

2 Assessment of Bridge Damage Research Based on Analytic Hierarchy Process

2.1 Time-Influenced Analytic Hierarchy Process

The long-span cable-stayed bridge structure is relatively complex. Many factors affect this bridge type, most of which cannot be expressed directly using functional expressions. This paper presents an analysis based on the time-influenced AHP.

Time-Dependent Reliability Theory and Calculation One of the more established methods for assessing the structural condition of long-span cable-stayed bridges, taking into account the influence of time on the structural assessment of bridges, is the time-dependent reliability theory [2]. Considering that the structure's resistance and load effects constantly change with time, the probability of failure also changes continuously. The time-dependent reliability of the system is shown in the following equation. Time-dependent reliability theory considers the effect of time dependence on reliability, which changes as the structural resistance gradually weakens over the structure's life.

Performance function of structure:

$$Z(t)=R(t)-S(t) \tag{1}$$

where R(t) is the structural resistance; S(t) is the load effect.

The probability of accomplishing a specific goal is as follows:

$$P_s(t)=P\left[R(t)-S(t)>0\right] \tag{2}$$

where $P_s(t)$ is the effective probability.

Changes in the resistance of structures over time are subject to many uncertainties, and the specific formula for calculating structural resistance is as follows:

$$R(t)=k_P r_P(t) \tag{3}$$

$$R_p(t)=R\left[f_{mi}(t),a_i(t)\right] \tag{4}$$

Where K_p is the random uncertainty variable; $R_p(t)$ is the calculated resistance of the structure; $F_{mi}(t)$ is the property of the ith material; $a_i(t)$ is the geometric parameter of the ith material.

Through document surveys, some scholars have carried out experimental analysis for structural and material resistance, statistically analyzed the law of continuous change of each parameter with time, and summarised the law of structural and material resistance with time as shown below [3]:

$$R(t)=R_0\varphi(t) \tag{5}$$

$$\varphi(t)=\begin{cases}1(0\le t\le t_c)\\1-w(t-20)(t\ge t_c)\end{cases} \tag{6}$$

Where R_0 is the structural resistance at t = 0, φ(t) is a statistically obtained deterministic function, and w is the rate of structural resistance decay.

Structural load effects are divided into structural dead load effects and structural variable load effects, further divided into varying load effects in general operation status and variable load effects in intensive operation status.

The load action model under dead load follows a normal distribution $N(1.0148W_k, 9.997^2)$, W_k is the standard deadweight value.

The specific calculation under variable load is shown in the following equation:

$$F_t(x)=\left\{\exp\left[-\alpha(x-u)\right]\right\} \tag{7}$$

$$F(x)=1-\exp\left[-(x/\eta)^m\right] \tag{8}$$

Where $F_t(x)$ is the load model for bridges in general operation status under variable load; F(x) is the load model for bridges in intensive operation status under variable load; α,μ,λ,m are correlation coefficients.

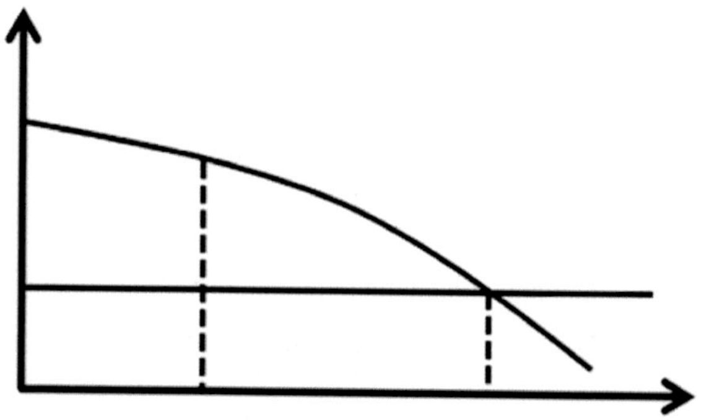

Fig. 1 Decay curve for structural reliability index

Structural time-dependent reliability index:

$$\beta(t) = \phi^{-1}\left[p_s(t)\right] \tag{9}$$

Where β(t) is the reliability index; ϕ^{-1} is the inverse function of the standard normal distribution function.

According to the previous research, the trend of the structural reliability index decaying with time is shown in Fig. 1. Finally, the corresponding reliability index is calculated by the structural time-dependent reliability method, based on which the reliability of the structure is judged.

Calculation of the Time Influence Coefficient The analysis of the time influence coefficient takes complete account of the changes in structural and material resistance with time. This paper refers to the method of time-dependent reliability. It determines the time influence coefficient for long-span cable-stayed bridge structures based on the change in the decay coefficient of structural resistance in time-dependent reliability to determine the final importance scale value, and the calculation process is shown below:

$$\mu = 1 + \left[1 - \varphi(t)\right] \tag{10}$$

$$\eta = \mu \times \eta_0 \tag{11}$$

Where φ(t) is the statistically obtained deterministic function; μ is the time influence coefficient; η_0 is the initial importance scale, and η is the importance scale after calculating the time influence.

According to the statistics, the resistance decay rate of bridge structures is w = 0.005. According to the code and some scholars' research, the resistance decay of the bridge structure in the first 20 years is negligible. The resistance decay coefficient function is shown in the following equation, and the trend is shown in Fig. 2.

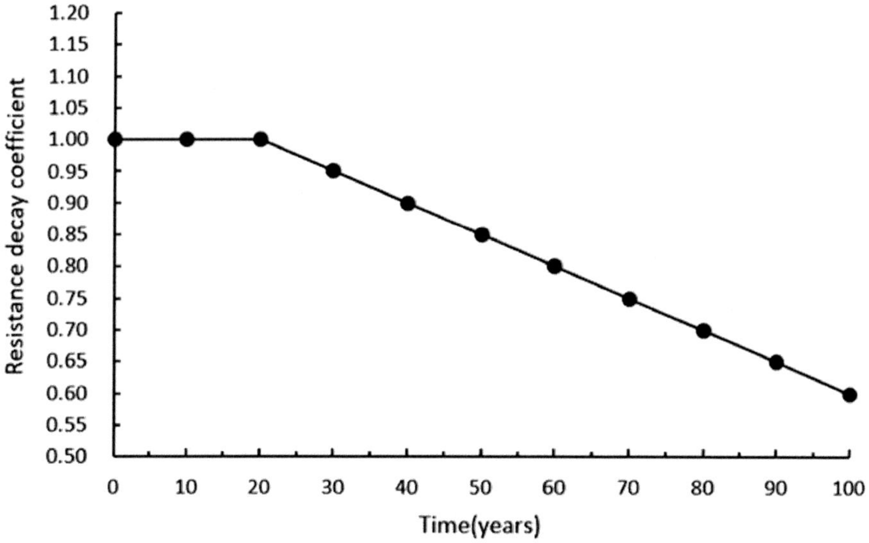

Fig. 2 Graph of structural resistance decay coefficient over time

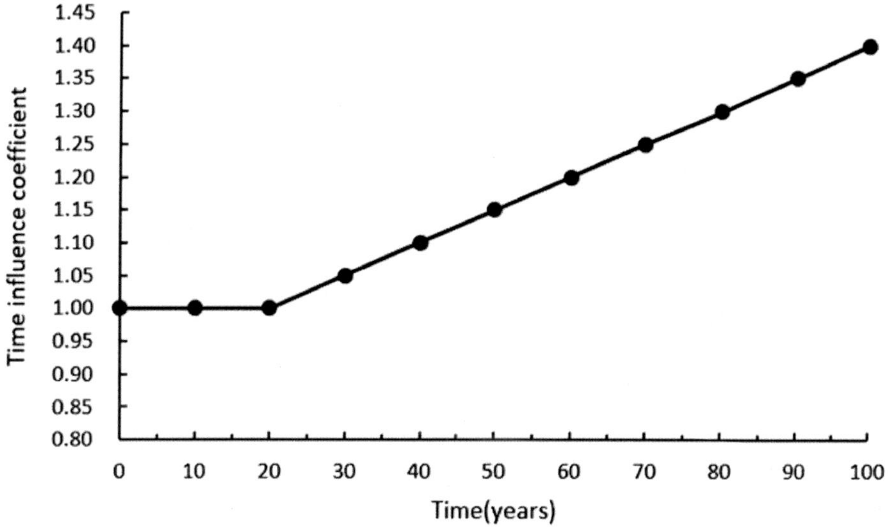

Fig. 3 Graph of change in time influence coefficient

$$\varphi(t)=\begin{cases}1(\leq t\leq 20)\\1-0.005(t-20)(t>20)\end{cases} \tag{12}$$

The corresponding time influence coefficient is determined according to the following equation. The specific changes in the coefficient are shown in the Fig. 3. It is clear from the graph that the time influence coefficient does not change significantly at the beginning of the completion of the bridge establishment. After 20 years, the coefficient shows a linear increase with time.

$$\mu = 1 + \left[1 - \varphi(t)\right] = \begin{cases} 1\left(0 \le t \le 20\right) \\ 1 + 0.005\left(t - 20\right)\left(t > 20\right) \end{cases} \tag{13}$$

2.2 Cable-Stayed Bridge Assessment Index System

Based on the AHP model [4], this paper adds a multi-scale analysis of the structure and a comprehensive condition assessment system for long-span cable-stayed bridges. This process was developed by evaluating long-span cable-stayed bridges at different scales from the model as a whole, components, elements, and welds. As shown in Fig. 4, the destination layer is the comprehensive condition assessment for long-span cable-stayed bridges, the component layer comprises five standard components, and the rest are the indicators of each element hierarchy. The affiliation between the various layers is clear at a glance.

2.3 Constructing Judgment Matrix and Consistency Check

In the AHP, each element a_{ij} in the judgment matrix represents a comparison of the degree of importance of any two indicators, i and j, in the same level of indicators to the previous level of indicators. To determine the differences in importance levels between indicators at the same layer, Satty et al [5] introduced the 1–9 scale method, where the specific size of a_{ij} is determined by using 1–9 and its reciprocal as a scale of importance levels. Table 1 specifies the importance scales between the indicators in the judgment matrix. Two indicators in the same layer are compared based on their impact on the previous layer, with 18 levels, 9 values, and 9 inverse of each value, indicating 18 different levels of importance. If a_{ij} is 3, it means that the impact on the upper layer, i indicator is 3 times that of j indicator and a_{ji} corresponds to a value of 1/3.

Based on the original scaling values of each weighted indicator at the same layer of the survey and the service life of the bridge, the corresponding time-influenced coefficient is calculated according to the specific time-influenced coefficient formula to work out the final scaling values between the indicators.

$$\eta = \mu \times \eta_0 = \begin{cases} \eta_0\left(0 \le t \le 20\right) \\ \eta_0\left[1 + 0.005\left(t - 2\right)\right]\left(t > 20\right) \end{cases} \tag{14}$$

This paper lists the time-influenced coefficient calculated every ten years based on the above formula. The specific values of the time-influenced coefficient up to one hundred years are shown in Table 2. As can be seen from Table 3, the time-influenced

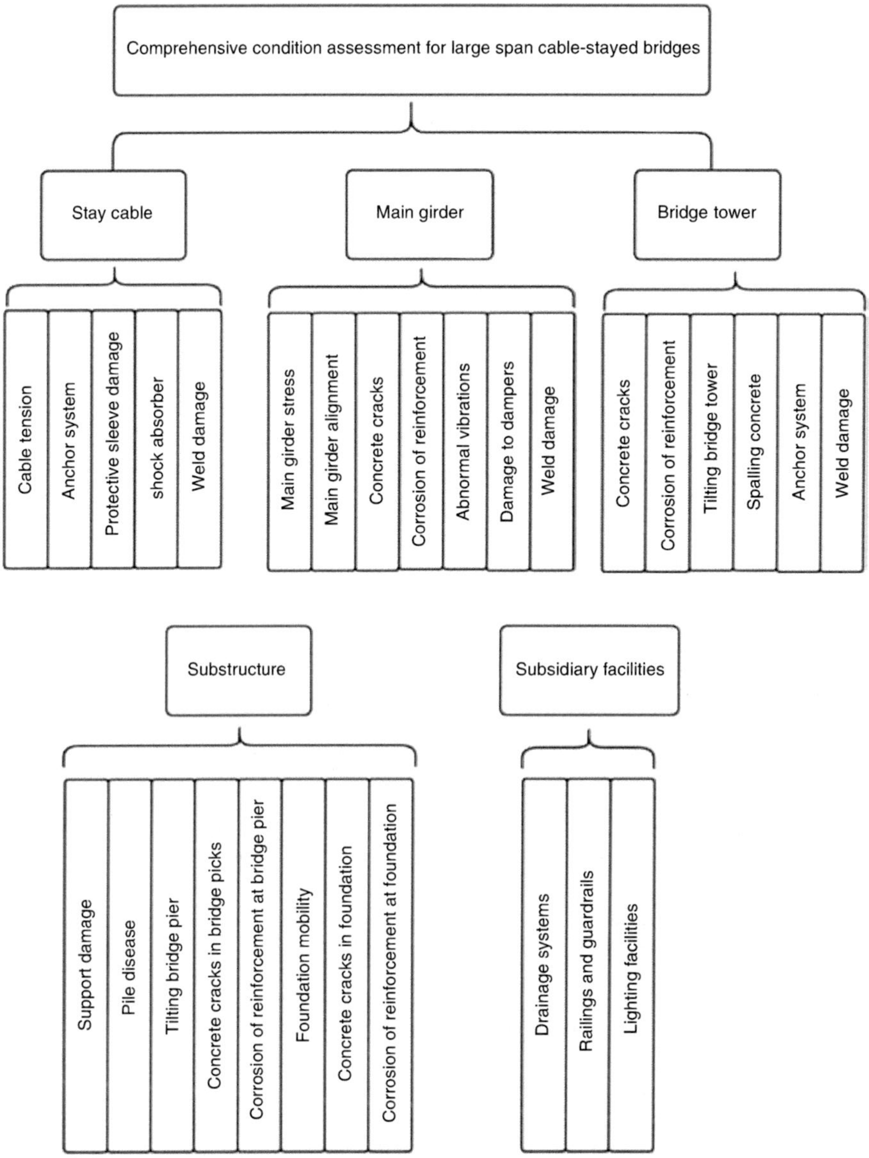

Fig. 4 Comprehensive condition assessment index system for long-span cable-stayed bridges

coefficient is 1.0 for the first 20 years from the start of the bridge's operational period. It changes to 1.4 by the end of the 100-year operating period, an increase of 40%, the magnitude of which cannot be ignored.

The final scale values of the judgment matrix are calculated from the values of the individual time-influenced coefficient. The specific values for each period are shown in Table 3. The scale values at this point are the scale values after considering the effect of time on the structure. There is a significant change in the scale when

Table 1 Initial importance scale η_0 of the judgment matrix

Importance Scale	Explanation
1	Indicates two indicators of equal importance
3	Indicates that the former indicator is slightly more important than the latter indicator when compared to the two indicators
5	Indicates that the former indicator is significantly more important than the latter indicator when compared to the two indicators
7	Indicates that the former indicator is more strongly important than the latter indicator when compared to the two indicators
9	Indicates that the former indicator is extremely more important than the latter indicator when compared to the two indicators
2,4,6,8	Indicates the middle value of the above importance discriminations
Reciprocal	Indicates the degree of importance of the two indicators as determined by the opposition ratio

Table 2 Time-influenced coefficient μ

Time (years)	20	30	40	50	60	70	80	90	100
Time-influenced coefficient μ	1.0	1.05	1.1	1.15	1.2	1.25	1.3	1.35	1.4

Table 3 Scales η after calculation of the time-influenced coefficient

	Time								
Scale	20 years	30 years	40 years	50 years	60 years	70 years	80 years	90 years	100 years
1	1.00	1.05	1.10	1.15	1.20	1.25	1.30	1.35	1.40
2	2.00	2.10	2.20	2.30	2.40	2.50	2.60	2.70	2.80
3	3.00	3.15	3.30	3.45	3.60	3.75	3.90	4.05	4.20
4	4.00	4.20	4.40	4.60	4.80	5.00	5.20	5.40	5.60
5	5.00	5.25	5.50	5.75	6.00	6.25	6.50	6.75	7.00
6	6.00	6.30	6.60	6.90	7.20	7.50	7.80	8.10	8.40
7	7.00	7.35	7.70	8.05	8.40	8.75	9.10	9.45	9.80
8	8.00	8.40	8.80	9.20	9.60	10.00	10.40	10.80	11.20
9	9.00	9.45	9.90	10.35	10.80	11.25	11.70	12.15	12.60

comparing values from 20 and 100 years, indicating a difference in the degree of importance of each indicator, reflecting the importance of the time-influenced coefficient on the bridge assessment. The specific values are determined based on the final scale values, and thus the judgment matrix is constructed.

Constructing the Judgment Matrix Each element a_{ij} in the judgment, the matrix represents the comparison of the importance of two indicators i and j, in the same layer of indicators as the previous layer and is constructed as matrix $A = (a_{ij})_{n \times n}$, called the judgment matrix. If the ratio of the importance of indicator i to indicator j is a_{ij}, then the ratio of indicator j to indicator i is $a_{ji} = \frac{1}{a_{ij}}$. By comparing all the indicators in the same layer two by two in this form, each layer is compared and calculated $\frac{n(n-1)}{2}$ times to finalize the values in the judgment matrix, which constitutes the judgment matrix. The eigenvector ω_0 corresponding to the largest eigenvalue λ_{max} of the judgment, the matrix is taken by comparing all indicators of each layer two by two through the judgment matrix. Then normalize, and the elements of the vector ω are the weights corresponding to the importance of each indicator at the same layer.

For example, the judgment matrix $A = (a_{ij})_{n \times n}$ consists of n interacting indicators at the same layer, and the maximum eigenvalue calculated corresponds to the eigenvector as follows:

$$B = [b_1 \, b_2 \, b_3 \cdots b_n] \tag{15}$$

Normalized vectors:

$$C = [c_1 \, c_2 \, c_3 \cdots c_n] \tag{16}$$

Where $C_i = \frac{|b_i|}{\sum_{i=1}^{n} |b_i|}; \sum_{i=1}^{n} c_i = 1$

Consistency Check *After calculating the weights for each indicator, the calculation results are verified for consistency. The three conditions for the judgment matrix to satisfy the consistency are as follows:*

1. *Diagonal elements are all 1.*
2. *Elements of symmetry about the diagonal are reciprocal.*
3. *Element priority transitive relationships are as follows:*

$$a_{ij} = \frac{a_{ik}}{a_{jk}}, i, j = 1, 2 \cdots, n, i \neq j \tag{17}$$

The judgment matrix allows for a quantitative and objective analysis to determine the weights of each indicator. However, to avoid non-consistency, the consistency check in the Analytic Hierarchy Process is used later to assess the reasonableness of the assessment results.

The consistency check formula is as follows:

$$CI = \frac{\lambda_{\max} - n}{n - 1} \tag{18}$$

$$CR = \frac{CI}{RI} \tag{19}$$

$$CR < 0.10 \tag{20}$$

Table 4 Average random consistency indicator value RI

n	1	2	3	4	5	6	7	8	9
RI	0	0	0.58	0.90	1.12	1.24	1.32	1.41	1.45

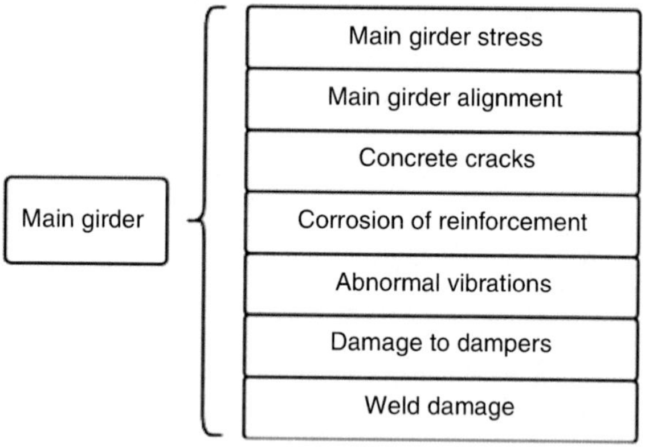

Fig. 5 Main girder assessment index system

Where *CI* is the consistency indicator; λ_{max} is the maximum eigenvector; *n* is the number of indicators; *CR* is the consistency ratio; and *RI* is the average random consistency indicator, for which Satty gives values corresponding to scales 1–9.

RI is a matrix formed by randomly selecting 500 sample data, constructing a judgment matrix by randomly selecting data from 1–9, taking the average of the maximum eigenvalue λ_{max} of each matrix, and then calculating the average random indicator value by the formula is shown in Table 4.

$$RI = \frac{\lambda_{\max} - n}{n - 1} \tag{21}$$

Determination of Indicator Weights Taking the seven indicators of the main girder in the structural state assessment of a long-span cable-stayed bridge as an example, Fig. 5 describes in detail the calculation process of the weights of each indicator.

We surveyed experts in the literature and assigned weighting values of indicators for the long-span cable-stayed bridge that introduces the influence of the indicators of the welding seam and determines the degree of importance of the indicators [6]. Then based on the current service life of the bridge, if a long-span cable-stayed bridge is currently in service for 10 years, calculate the time-influenced coefficient $\mu = 1.0$ according to the formula, and determine the scaled value η of the judgment matrix after accounting for time as 1.00, 2.00, 3.00, 4.00, 5.00, 6.00, 7.00, 8.00, 9.00. Using these nine scales to determine the proportion of importance of each indicator in the judgment matrix for a two-by-two comparison, then determine the judgment matrix $A = (a_{ij})_{7 \times 7}$of the main girder assessment indicators for the condition assessment of long-span cable-stayed bridges, as shown in the figure below, the judgment matrix is a 7-row, 7-column matrix, and the importance ratio between the two indicators is determined from the table.

$$A = \left(a_{ij}\right)_{7\times 7} = \begin{bmatrix} a_{1,1} & \cdots & a_{1,7} \\ \vdots & \ddots & \vdots \\ a_{7,1} & \cdots & a_{7,7} \end{bmatrix}$$
$$= \begin{bmatrix} 1.000 & 0.250 & 0.500 & 0.333 & 3.000 & 3.000 & 2.000 \\ 4.000 & 1.000 & 3.000 & 2.000 & 6.000 & 6.000 & 5.000 \\ 2.000 & 0.333 & 1.000 & 0.500 & 4.000 & 4.000 & 3.000 \\ 3.000 & 0.500 & 2.000 & 1.000 & 5.000 & 5.000 & 4.000 \\ 0.333 & 0.167 & 0.250 & 0.200 & 1.000 & 1.000 & 0.500 \\ 0.333 & 0.167 & 0.250 & 0.200 & 1.000 & 1.000 & 0.500 \\ 0.500 & 0.200 & 0.333 & 0.250 & 2.000 & 2.000 & 1.000 \end{bmatrix} \tag{22}$$

With the calculated judgment matrix A, the maximum eigenvalue $\lambda_{max} = 7.145$ corresponding to the matrix is calculated, thus determining the eigenvector as follows:

$$\omega_0 = \begin{bmatrix} -0.2141 & -0.7462 & -0.3312 & -0.5037 & -0.087 & -0.087 & -0.1368 \end{bmatrix} \tag{23}$$

The eigenvector is then vector-normalized, and the normalized eigenvector value is calculated and found:

$$\omega = \begin{bmatrix} 0.102 & 0.354 & 0.157 & 0.239 & 0.041 & 0.041 & 0.065 \end{bmatrix} \tag{24}$$

Finally, according to the formula, perform a consistency check on the calculation result of the judgment matrix and find the consistency index.

$$CI = \frac{\lambda_{\max} - n}{n-1} = \frac{7.145-7}{7-1} = 0.02417 \tag{25}$$

The proportionality check for consistency is as follows:

$$CR = \frac{CI}{RI} = \frac{0.02417}{1.32} = 0.0183 < 0.10 \tag{26}$$

The results of the above formulae show that the main girder assessment index for the comprehensive condition assessment of long-span cable-stayed bridges meets the consistency requirements. Calculating the judgment matrix and the consistency check, the corresponding weight values for each indicator of the main girder at the 10-year service life of the cable-stayed bridge were finally determined, as shown in Table 5.

Based on the above calculations, the weights of each hierarchical status evaluation indicator for long-span cable-stayed bridges can be determined for each time point in their service life. We set the weight values of each indicator corresponding to the 20-year (Table 6) and 50-year (Table 7) time points for long-span cable-stayed bridges.

It is clear from Table 7 that the weight of each indicator in the long-span cable-stayed bridge index system changes over time, taking into account the influence of

Table 5 Weighting values for each indicator for the main girder

Main girder indicators	Weights
Main girder stress	0.102
Main girder alignment	0.354
Concrete cracks	0.157
Corrosion of reinforcement	0.239
Abnormal vibrations	0.041
Damage to dampers	0.041
Weld damage	0.065

Table 6 The weighting of indicators for comprehensive condition assessment of long-span cable-stayed bridges (t = 20 years)

Component indicators	Weights	Element indicators	Weights
Stay cable	0.109	Cable tension	0.419
		Anchor system	0.263
		Protective sleeve damage	0.062
		Shock absorber	0.097
		Weld damage	0.160
Main girder	0.184	Main girder stress	0.102
		Main girder alignment	0.354
		Concrete cracks	0.157
		Corrosion of reinforcement	0.239
		Damage to dampers	0.041
		Weld damage	0.065
Bridge tower	0.319	Concrete cracks	0.111
		Corrosion of reinforcement	0.198
		Tilting bridge tower	0.335
		Spalling concrete	0.198
		Anchor system	0.111
		Weld damage	0.047
Substructure	0.319	Support damage	0.051
		Pile disease	0.034
		Tilting bridge pier	0.235
		Concrete cracks in bridge piers	0.082
		Corrosion of reinforcement at bridge pier	0.140
		Foundation mobility	0.235
		Corrosion of reinforcement at foundation	0.140
Subsidiary facilities	0.068	Drainage systems	0.400
		Railings and guardrails	0.400
		Lighting facilities	0.200

different time points. Table 8 shows the weighting of indicators at the beginning, middle, and end of the operation of long-span cable-stayed bridges at 20 and 50 years and also shows an apparent increase and decrease in the weighting of the indicators at the two times. The weight changes for each indicator of the main girder

Table 7 The weighting of indicators for comprehensive condition assessment of long-span cable-stayed bridges (t = 50 years)

Component indicators	Weights	Element indicators	Weights
Stay cable	0.099	Cable tension	0.445
		Anchor system	0.263
		Protective sleeve damage	0.053
		Shock absorber	0.087
		Weld damage	0.152
Main girder	0.177	Main girder stress	0.098
		Main girder alignment	0.377
		Concrete cracks	0.154
		Corrosion of reinforcement	0.244
		Damage to dampers	0.035
		Weld damage	0.059
Bridge tower	0.332	Concrete cracks	0.102
		Corrosion of reinforcement	0.198
		Tilting bridge tower	0.361
		Spalling concrete	0.198
		Anchor system	0.102
		Weld damage	0.041
Substructure	0.332	Support damage	0.045
		Pile disease	0.029
		Tilting bridge pier	0.249
		Concrete cracks in bridge piers	0.076
		Corrosion of reinforcement at bridge pier	0.138
		Foundation mobility	0.249
		Corrosion of reinforcement at foundation	0.138
Subsidiary facilities	0.060	Drainage systems	0.411
		Railings and guardrails	0.411
		Lighting facilities	0.178

Table 8 Value of the changes in the weight of each indicator for the main girder

Each indicator for the main girder	Weighting of 20 years	Weighting of 50 years	Percentage difference (%)
Main girder stress	0.102	0.087	−14.71
Main girder alignment	0.354	0.410	15.82
Concrete cracks	0.157	0.148	−5.73
Corrosion of reinforcement	0.239	0.249	4.18
Abnormal vibrations	0.041	0.028	−31.71
Damage to dampers	0.041	0.028	−31.71
Weld damage	0.065	0.051	−21.54

at the 20-year and 50-year time points are presented as examples (Table 8). The weight of the linear indicator in the main girder increased by 15.82%, and the weight of the reinforcement indicator increased by 4.18% (Table 8). This suggests that these two indicators have a greater weight and are relatively more important. The remaining indicators have decreased to varying degrees, with the vibration and damper damage indicators decreasing by 31.71%, indicating the indicators have a relatively small weighting and a slightly weaker degree of importance.

2.4 Synthetic Calculation Method for Long-Span Cable-Stayed Bridges

According to the standard of China's "Specification For Highway Bridge Culvert Maintenance" (JTGH11–2021) [7], the calculation method of the damage condition index is used for the synthetic calculation of long-span cable-stayed bridges. The corresponding deduction criteria are established for each element of long-span cable-stayed bridges. Under the determined weights, the formula is used to find the scoring value for each component and determine its component level according to the criteria developed. Then, based on the component weights and formula, the condition score of the entire bridge is solved, and the full bridge rating criteria determine the whole bridge rating level.

$$D_j = 100 \times \left(1 - \sum_{i=1}^{n} R_i W_i / 5\right) \tag{27}$$

Where D_j is the score obtained for the jth component; R_i is the rating deduction scale for the ith indicator; and W_i is the weight of the ith indicator.

$$D_r = \sum_{j=1}^{n} D_j W_j \tag{28}$$

Based on the classification and assessment criteria in the "Specification For Highway Bridge Culvert Maintenance" (JTGH11–2021) [8], we set the deduction evaluation criteria for bridge element indicators on a scale of 0–5, indicating six states from intact to dangerous for each element, corresponding to five levels (Table 9). For long-span cable-stayed bridges, the results of the comprehensive condition assessment were divided into five classes (Class I, II, III, IV, V) (Table 10).

Table 9 The deduction evaluation criteria for long-span cable-stayed bridge elements

Class ratings	Score range	Status description
Class I	$88 \le D_r \le 100$[a]	Intact condition
Class II	$60 \le D_r \le 88$	Minor damage, good condition
Class III	$40 \le D_r \le 60$	There is some damage, which has an impact on the function of the bridge
Class IV	$D_r < 40$	Significant damage exists, which seriously affects the regular use of the bridge

[a] D_r is the full bridge and individual component indicator score

Table 10 Component and overall rating of long-span cable-stayed bridges

Element deduction values	Status description	Corresponding classes
0	Intact	Class I
1	Good	Class I
2	Better	Class II
3	Poor	Class III
4	Very poor	Class IV
5	Dangers	Class V

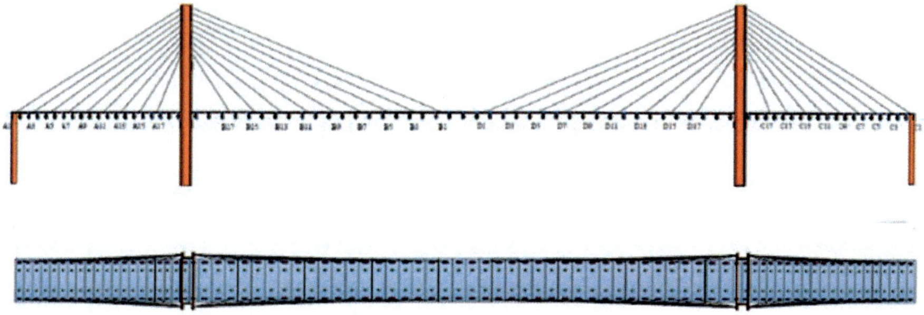

Fig. 6 Schematic diagram of the cable-stayed bridge structure

3 Example Analysis

3.1 Validation Experiments Based on Experimental Models of Bridges

As it is impossible to conduct damage tests on real bridges, this project took the alternative route of using the ANSYS bridge model for data collection.

ANSYS Bridge Model

Finite element modeling. In physical model tests, many conditions cannot be achieved ideally, such as various types of damage conditions. Therefore, it is necessary to build a finite element model according to the physical test model parameters and analyze it numerically. So we use the structural finite element software ANSYS to construct a finite element model of the model structure and analyze it numerically (Fig. 6).

Classification of Bridge Levels and Deterioration The technical condition assessment of the bridge is broken down into four groups: full bridge, four structural parts, components, and elements (Fig. 7 [9]. In particular, elements such as girder, pier body, top cap, pile cap, and pile foundation were used as the lowest level assessment unit, and an element disease database was established according to the characteristics of the individual element, which will be graded according to the degree of disease [10].

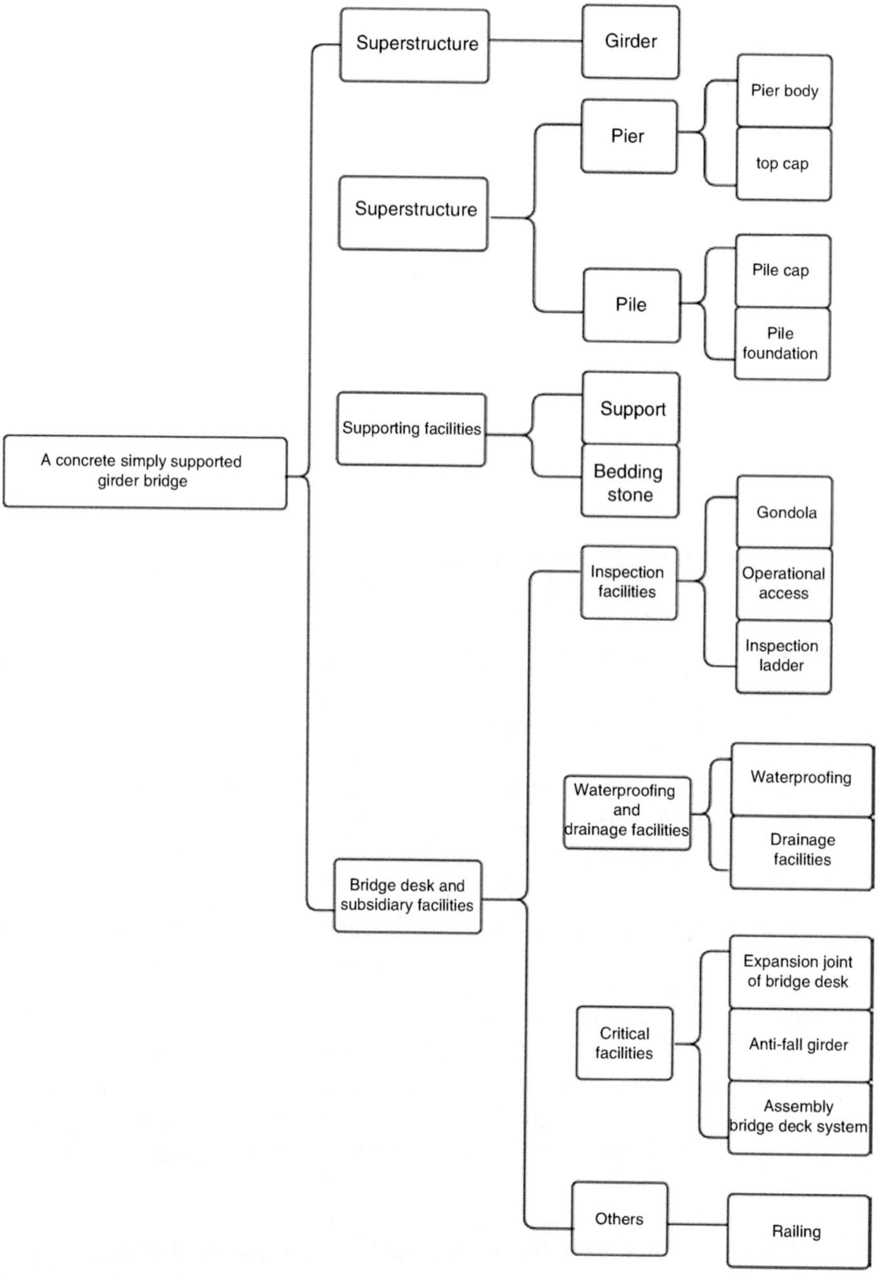

Fig. 7 Composition of the bridge

Calculation of weights. Based on the AHP, a hierarchical structure of this bridge was established (Table 11). The target layer is the assessment of the technical condition of the whole bridge, the criterion layer mainly includes the superstructure,

Table 11 Structure of the bridge rating system

Target layer	Criterion layer	Index layer
Full bridge technical condition assessment	Superstructure	Girder
	Substructure	Pier
		Pile
	Supporting facilities	Support
		Bedding stone
	Bridge deck and subsidiary facilities	Inspection facilities
		Waterproofing and drainage facilities
		Critical facilities
		Others

Table 12 Weighting of the four major structural parts

Bridge parts	Superstructure	Substructure	Supporting facilities	Bridge deck and subsidiary facilities
Weights	0.4103	0.4103	0.1220	0.0574

substructure, supporting facilities, bridge deck, and subsidiary facilities. Under the criterion layer, several subindexes are set separately, forming the index layer [11].

Take the weighting analysis at the criterion layer as an example, calculate the weights attributed to each indicator, and build a judgment matrix:

$$U = (u1,\quad u2,\quad u3,\quad u4) \tag{29}$$

Where U is the assessment of the technical condition of the whole bridge; $u1$ is the superstructure; $u2$ is the substructure; $u3$ is the supporting facilities; $u4$ is the bridge deck and subsidiary facilities. Based on the 0–2 triple scale method, the relative importance of each index is determined, and the importance matrix was established:

$$U = \begin{bmatrix} & u_1 & u_2 & u_3 & u_4 \\ u_1 & 1 & 1 & 2 & 2 \\ u_2 & 1 & 1 & 2 & 2 \\ u_3 & 0 & 0 & 1 & 2 \\ u_4 & 0 & 0 & 0 & 1 \end{bmatrix} \tag{30}$$

The weights of the four structural parts calculated using the AHP based on the importance matrix are shown in Table 12. As can be seen, the superstructure and substructure are given greater weight, the support facilities are second, and the bridge deck and ancillary facilities have the least weight.

Similarly, the superstructure has only the girder part of the remaining components, which therefore has a weight of 1. The substructure has two elements, the pier, and the pile, which are equally important (weight = 0.50). The support area also consists of 2 elements: the support and the bedding stone. Each element is of equal importance, with a weight of 0. 50. There are four elements in the structural part of the bridge deck and subsidiary facilities. The weights of each component are in Table 13.

Technical Condition Assessment of the Bridge The information on the degradation present on the bridge through manual inspection is shown in Table 14 [12].

Table 13 Weighting of structural parts of the bridge deck and subsidiary facilities

Bridge parts	waterproofing and drainage facilities	Inspection facilities	Critical facilities	Others
Weights	0.1798	0.1798	0.5850	0.0553

Table 14 Types of disease to the bridge

Structural parts	Components	Elements	Damage types	Degradation level
Superstructure	Girder	Main girder	Fine reticulated cracks in the concrete surface	C
Substructure	Pier	Pier body	Fine cracks, width within 0.3 mm	C
Supporting facilities	Support	Steel parts	Microcracking of steel parts	C
Bridge deck and subsidiary facilities	Waterproofing and drainage facilities	Bridge deck waterproofing	Localized blistering, peeling, and cracking of the waterproofing layer	C
	Inspection facilities	Operational access	Missing or damaged step plate	B
	Critical facilities	Protective wall	Exposed concrete, falling blocks	B

Table 15 Scoring of bridge structures

Elements	Score	Components	Score	Structural parts	Score	Full bridge score
Bridge deck waterproofing	90.0	Waterproofing and drainage facilities	82.5	Bridge deck and subsidiary facilities	91.7	93.8
Waterproof facilities	75.0					
Operational access	80.0	Inspection facilities	93.3			
Protective wall	80.0	Critical facilities	93.3			
Main girder	90.0	Girder	90.0	Superstructure	90.0	
Pier body	90.0	Pier	95.0	Substructure	97.5	
Steel parts	90.0	Support	90.0	Supporting facilities	95.0	

Based on the above calculation steps, the scoring of elements, components, structural parts, and the entire bridge is in Table 15. The table shows that the bridge has the lowest waterproofing and drainage facilities score and should be inspected more frequently. The bridge deck and subsidiary facilities, superstructure health code, is 8, substructure and support facilities health code is 9, and the whole bridge health code is 8, indicating that all structural parts of the bridge and the entire bridge are currently in very good condition.

The assessment of the bridge condition using this method not only visualizes the state and health of the elements but also provides a real-time overview of the operational condition of every component by scoring each part of the bridge. This technique facilitates the development of maintenance and repair plans for the bridge structure. Scoring the whole bridge gives an idea of the overall operational status of the bridge. Compared to other assessment methods, this method combines the AHP to provide a quantitative description of the bridge's condition.

4 Conclusion

This paper proposes a time-influenced AHP, which determines the weights of each evaluation index, improves the efficiency of a bridge inspection, and reduces work bias. A finite element model of bridges was also developed using ANSYS to validate the AHP time influence. This paper establishes a method for assessing the technical condition of bridges based on the AHP. Using bridge deterioration as the basis for developing point deduction rules focuses on the importance of parts of bridges. It uses a combination of qualitative and quantitative analysis to determine the weights of each component and structural part of the bridge structure.

Based on the AHP, the bridge technical condition assessment method draws on the relevant analysis ideas of bridge assessment at home and abroad [13]. It combines with the "Specification For Highway Bridge Culvert Maintenance" (JTGH11-2021) to assess the bridge level by level, which can grasp the structural state of the bridge from the detail to the whole and facilitate bridge management. The example calculation shows that the method provides a visual overview of the bridge disease and a real-time assessment of the bridge's structural condition and provides timely guidance for inspection, maintenance, and repair.

References

1. Huang ZW, Huang Q, Ren Y. A method for assessing small and medium-sized bridges based on degree analysis. Journal of Southeast University (JSEU_EE), 42(2): 164–168, (2012).
2. Wang Z. Bridge technical condition assessment based on Fuzzy analytic hierarchy process. Transportation Science & Technology, (5): 68–71, (2016).
3. Liu HB, Wang XQ, Jiao YB, et al. Condition Evaluation for Existing Reinforced Concrete Bridge Superstructure Using Fuzzy Clustering Improved by Particle Swarm Optimization. Structure and Infrastructure Engineering, 13(7): 955–965, (2017).

4. Huang Q, Ren Y, Lin Y. Application of Uncertain Type of AHP to Condition Assessment of Cable-stayed Bridges. Journal of Southeast University (English Edition), 23 (4):599–603, (2007).
5. Ma Qingyue. Research on comprehensive evaluation technology of combat effectiveness of weapon and equipment system[D]. Harbin Institute of Technology, (2015).
6. Huang Q, Tang HH, Ren Y, et al. Research on the evaluation theory of long-span bridges based on fuzzy theory. Journal of Highway and Transportation Research and Development, 27(1): 62-66, (2010).
7. Ministry of Transport of the People's Republic of China. Specification For Highway Bridge Culvert Maintenance: JTG H11—2004. Beijing: People's Transportation Press, (2004).
8. The Ministry of Construction of the People's Republic of China. Technical standard of maintenance for city bridge: CJJ 99—2003. Beijing: China Architecture & Building Press, (2003).
9. Ministry of Transport of the People's Republic of China. Standards for technical condition evaluation of highway bridges: JTG/T H21—2011. Beijing: People's Transportation Press, (2001).
10. The Ministry of Railway of the People's Republic of China. Evaluation Standard for Degradation of Railway Bridge and Tunnel buildings: TB/T 2820.1—2820.8.Beijing: China Railway Publishing House, (1998).
11. The Ministry of Railway of the People's Republic of China. Repair rules for railway bridge and tunnel buildings: Railway Transport [2010] No. 38. Beijing: China Railway Publishing House, (2010).
12. The Ministry of Railway of the People's Republic of China. Rules for the repair of high speed railway bridge and tunnel buildings: Railway Transport [2011] No. 131. Beijing: China Railway Publishing House, (2011).
13. Huang HY. Comprehensive assessment of operational risks of in-service suspension bridges and practical application research. Kunming: Kunming University of Science And Technology, (2018).

Hill-Patch System as a Strategic Management Model to Activate and Conserve Urban Axes: The Case of the San Bernardo Green Corridor, Santiago, Chile

Loreto Rudolph, Stefania Pareti, Vicente Valdebenito, and Samuel Carmona

Abstract

This study explores how the patch-hill system works as a strategic model to activate and preserve urban axes located in peripheral areas of the city, mediating between urban and rural areas. The Green Corridor of San Bernardo in Santiago de Chile has been selected as a case study, due to (1) the opportunities present in the peri-urban geography of the city to harmonize the ecological-urban imbalances, (2) it is a set of five hills interconnected by a network of canals in a territory in the process of urban expansion, (3) functioning as a transition area on a metropolitan and communal scale and (4) reducing the erosive impact of the city, and increasing recreational and productive areas and promoting the mobility of the local fauna. The methodology consists of an analysis of the phenomenon of territorial expansion in San Bernardo. Then, the relationship between the project and the hills through which it passes is studied, with emphasis on the agricultural proposal of the Hasbún hill. We conclude that constructing a border that generates relationships between city and hill contributes to the conservation and activation of peripheral urban axes, mediating between the urban and rural, contributing to the consolidation of sustainable cities.

L. Rudolph (✉) · V. Valdebenito · S. Carmona
Universidad Técnica Federico Santa María, Valparaíso, Chile
e-mail: loreto.rudolph@sansano.usm.cl; vicente.valdebenito@sansano.usm.cl; samuel.carmona@sansano.usm.cl

S. Pareti
Universidad Andrés Bello, Santiago, Chile
e-mail: s.pareti@uandresbello.edu

J. T. Anderson (ed.), *Advances in Environmental Sustainability*, Springer Proceedings in Earth and Environmental Sciences,
https://doi.org/10.1007/978-3-031-26365-1_13

Keywords

Green corridor · Sustainable cities · Urban limits

1 Introduction

1.1 A Subsection Sample

This research aims to study the functioning of the proposed Integrated Patch System, winner of the Contest for the Cerros Isla project for the San Bernardo Green Corridor as a model to activate and conserve urban axes. This project is located in the San Bernardo commune, which belongs to the capital of Santiago and is located in the southwestern part of it (Fig. 1). This commune is considered an urban limit of the capital since it currently divides the metropolitan area from the rural one. In a context of constant urbanization, the urban limits have changed, understanding the rural spaces bordering the metro area as adrift sites and their edges as a rigid limit and divider of a landscape unit [1]. The contest carried out by the Catholic University of Chile and the Santiago Cerros Isla Foundation contributes to the integration of the landscape in territorial planning and the design of public spaces, through an ecosystem approach that recognizes the opportunities present in the geography of Santiago for harmonizing ecological-urban imbalances in the peri-urban area of the city [2].

Fig. 1 Comune of San Bernardo Ubicated in Santiago city. (Own Elaboration)

These two types of spaces with different characteristics are being adjusted through a mediating project.

The proposal for the Green Corridor of the commune of San Bernardo exposes an emblematic case, as it is a set of five interconnected hills near the Maipo River - Chena, Hasbún, Negro, Adasme, and Los Morros- by a network of canals that make up a potential green corridor (Fig. 2) in a territory in the process of urban expansion [3]. This project presents an intervention strategy based on sustainable patches of land, at strategic points linked to the geographical landmarks mentioned above. It is open to long-term development in. It allows parallel financing methods. and administration for each one, to incorporate in the design the times and actors neighborhood, municipal and regional management. These patches seek to build a symbiotic relationship between each hill and the city [4].

According to the Metropolitan Regulator Plan of Santiago (PRMS100), Santiago will present substantial urban growth that would directly impact San Bernardo's commune, [5]. From this perspective, San Bernardo will be involved in an urbanization process that would reduce its productive areas and isolate its territorial elements, such as its hills and the Maipo River. San Bernardo has the opportunity to act as a precedent in constructing a new transitional boundary between the countryside and the city, [4], using its natural landmarks in conjunction with an urban project to propose a transition area on a metropolitan and communal scale between both worlds. Simultaneously it provides a corridor of territorial elements that reconstruct

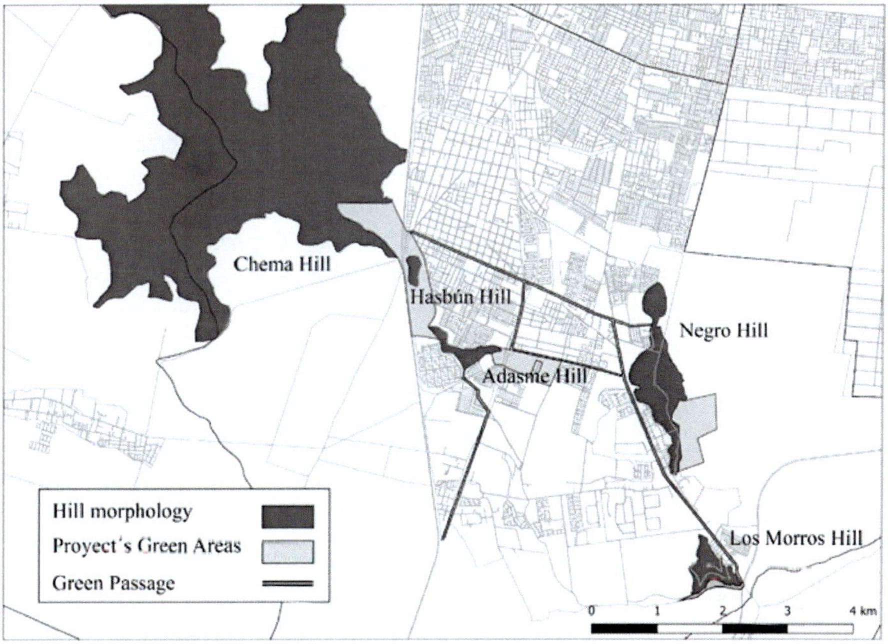

Fig. 2 Zones of intervention of winner's proposal. (Own Elaboration)

and reformulate the experience of the rural and productive land-scape of the central valley.

2 Theoretical Framework

The theoretical framework of this research focuses on explaining the concepts of urban boundaries, green corridors, and sustainable cities.

Cities are becoming increasingly complex systems from social, economic, and ecological points of view [6]. The continuous expansion of the urban fabric, both in territorial extension and population density, gives rise to growing demands on the social infrastructure to meet everyday needs. Due to the capital's growing housing and product demand, it is necessary to expand its limits, which leads to a constantly renewed transition between the urban-built and the rural-productive. San Bernardo is in this condition, known as a peri-urban area. Peri-urban areas are the transition, or interaction zones where urban and rural activities are juxtaposed. Landscape characteristics are subject to rapid modifications induced by human activities [7] and the city's expansion. In this context, the availability of an adequate urban transportation infrastructure that allows for the movement of people and goods in a dignified manner [8], while connecting the different productive spheres of the region, is fundamental.

In the discussion of the conservation and activation of urban spaces in a context that borders the rural, the idea of a green corridor implemented in one of the largest housing communities of the city speaks of an intermediary logic. Green or sustainable architecture has to do with sustainability and is related to design rather than a statistical indicator [9]. Green transport corridors represent trans-shipment routes with a concentration of freight traffic between major hubs and long transport distances marked by reduced environmental and climate impact [10].

In the context of a harmonized transport system, the green corridor concept represents a cornerstone in developing and implementing integrated and sustainable transport solutions, [11]. The spatial-explicit and ecological connectivity analyses of urban green infrastructure are helpful tools for planning and evaluating spatial patterns and their changes for the sustainability of urban development [12].

In the case of San Bernardo, urban expansion over rural areas and the transformation of productive spaces into urbanized areas have fragmented the recognition of a territorial identity of the fertile landscape of the central valley [3]. For this reason, the project proposes a system of recreational-productive parks in the form of sustainable green patches with a design strategy based on replicability, so that the design can be brought to reality progressively and systematically within an ever-expanding urban fabric.

One of the important essential properties of green corridors is the involvement of public and private stakeholders, including the political level, requiring new governance models to manage green corridors. [12]. At the same time, the corridors play a role in social and neighborhood activation, proposing places of relaxation and recreation to the community. Green corridors connecting isolated remnant habitat patches can increase connectivity and provide city ecosystem services. Vacant land, especially prevalent in shrinking cities, presents a unique opportunity to reconnect these landscapes [13].

Medellin is an example of successful green corridors in the Latin American context, creating 30 green corridors that connect the city with a sustainable approach under energy conservation and cooling through planting extensive linear streets [14]. Another example of green corridors is Cheshire in northwest England, with an area of 72 hectares, equivalent perhaps to a large urban park. However, its extensive perimeter of over 38 km gives rise to quite clear liabilities concerning maintenance, particularly fencing. This case suggests that "green corridors"can play a strategic role as traffic-free recreational routes and local open spaces [13]. Singapore's last and most similar case is a highly urbanized city, demanding housing construction and with few green areas available. The corridor implementation, in this case, aimed to improve the lives of people in the city through an island-wide network of park connectors [15].

3 Methodology

The methodology of this study consists of the survey and territorial analysis of the hills of the commune of San Bernardo through which the Parches-Cerro project passes. These are the Chena, Negro, Asme, Los Morros, Río Maipo, and Hasbún Hills; the latter will be analyzed in depth, exploring the Patch program and its relationship with the immediate context.

In the first place, a survey of information was carried out regarding the population growth that the commune of San Bernardo experienced in 1993, 2003, and 2020 to the present, evidencing that population growth is generated towards the periphery of the urbanization, seeing each time plus the limit with the most diffuse rural. Secondly, we evaluated how the urban-rural boundary established by the Santiago Regulatory Plan was related to the streets of the San Bernardo commune. Finally, a specific hill was analyzed; the Patch-Hill Hasbún was chosen because it is located in the urban limit and houses programs in the project activators of localities. For this, an analysis of the Project Plan was carried out together with an analysis of the surrounding programs, and transport axes, among other factors that can be decisive in identifying the Hasbún Patch as an urban activator.

4 Results

The urban expansion in Santiago de Chile has transformed production spaces towards urban areas, assuming a condition of the encounter between both urban-rural contexts. As seen in the survey of satellite photographs in Figs. 3, 4, and 5, the limit between the urban and rural areas has decreased over the years, extending the city, which has brought various adverse consequences, but it can also mean an opportunity. To enhance the sense of local identity by creating dynamic spaces that

Figs. 3, 4, and 5 From left to right. Satellite images 1993, 2003, and 2020 of the commune of San Bernardo. (Google Earth)

Fig. 6 San Bernardo Commune Map. Urban Limit

establish an interaction between the city and the rural through a limit generated as programmatic parks that are simultaneously recreational and productive.

Likewise, the territorial survey of Figs. 6 and 7 shows that the population of the commune of San Bernardo is concentrated to the north, being established as the urban sector of the commune, while the rural sector is to the south. According to the Regulatory Plan for the Santiago Metropolitan Region (PRMS100), significant urban growth will directly impact the commune of San Bernardo, which is understood as an urban-rural boundary of the city.

The proposed Patch-Cerro system covers the five hills that are part of the commune of San Bernardo; these are located outside the urban sector of the commune, (Fig. 8), which defines the limit between the urban and the rural.

Hasbún Hill is adjacent to the central highway and framed by Enrique Madrid Osorio, Eyzaguirre, and General Urrutia streets. Towards the east side of the mountain is the city of the commune, it is mainly a residential neighborhood, with the

Fig. 7 San Bernardo Commune Map San Bernardo commune streets

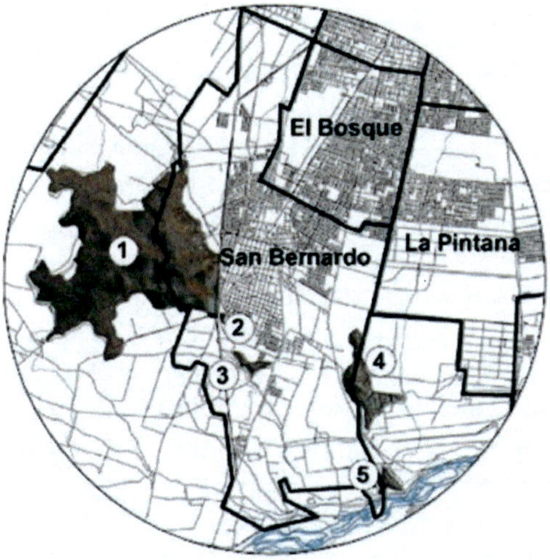

Fig. 8 San Bernardo commune map: Hills integrated into the green corridor. (Own Elaboration)

presence of schools, commercial premises, and several bus stops along Mateo de Toro and Zambrano streets.

A public transport route close to the hill is vital activation factor for the axis, since it allows the patch to be accessible using public transport.

Hasbún Hill is located adjacent to the central highway, and is framed by Enrique Madrid Osorio.

The Hasbún Patch proposes creating a a meeting space for the different focuses of agricultural culture, whether local cooperatives, companies, technical training

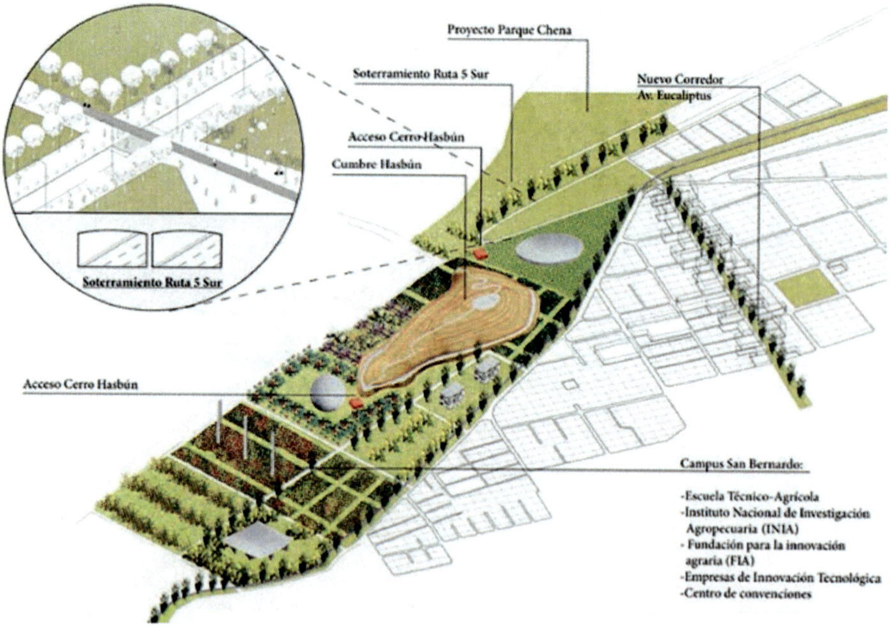

Fig. 9 Assembly drawing. Hasbun patch. (Map: Vasquez, Samaniego, Walker, Mateluna, Salas)

centers, or other institutions. A development and research campus is proposed that provides adequate infrastructure for the various institutions (Fig. 9).

This patch seeks to build a relationship between the hill and the city, integrating available services related to agriculture that function as mediators between the immediate community and the territorial elements that are sought to be preserved.

Among the strategies that stand out in the Patch is the enabling of observation infrastructure that allows the visualization of the summits of the hills of the Central Valley; in addition, landscape and ecological attributes are preserved to conserve native flora and fauna of the Central Valley, a gradual restoration accompanied by educational infrastructure, valuing the ecology of the sector.

5 Conclusion

Due to the condition of the commune of San Bernardo to expand, growing to-wards the periphery, the opportunity is generated to project green axes to activate urban sectors, acting as a precedent in the construction of a new limit between the countryside and the city. As a transition between both worlds, it is reconstructing and reformulating at the same time the experience of the rural and productive landscape together with the dynamics of urbanization in a public way and open to all citizens. In the study, a survey of territorial information was carried out, where it was concluded that the hill patch system proposed for the commune of San Bernardo is an activating factor of urban planning because it establishes the relationship between the urban and rural limits through strategies focused on the enabling of infrastructure that allows public access, generating meeting and learning spaces through the integration of both contexts: urban and rural.

References

1. Laandum 2019 Integrated Patch System
2. Cerros Isla Foundation, 2015 San Bernardo Green Corridor
3. Urban Platform, 2015 "Cerros Isla Contest. Green Corridor of San Bernardo" announces the winning project
4. Walker F, Vásquez A, Samaniego J, Mateluna L, Fonck M, Rojas S, Salas F, Chateau F and Burdiles M 2017 Integrated Patch System, reformulate the experience of the productive landscape in San Bernardo
5. Chilean Chamber of Construction 2014 Aprobation PRMS100; Analysis and scopes.
6. Dietz T, Carpenter S, Alberti R, Folke M and Moran C 2007 Complexity of coupled human and natural systems.
7. Douglas I 2006 Peri-Urban Ecosystems and Societies: Transitional Zones and Contrasting Values.
8. Sanchez R and Lupano J 2008 Movilidad Urbana, Políticas de movilidad urbana e infraestructura urbana de transporte
9. Xiao F and Wei Y 2018 Basic Green Building Design: Reconnecting Sustainability to the Vernacular
10. Prause G 2014 A Green Corridor Balanced Scorecard
11. Hunke K and Prause G 2013 Management of Green Corridor Performance
12. Wang H and Zongping P 2020 Urban Green Corridors Analysis for a Rapid Urbanization City Exemplified in Gaoyou City, Jiangsu
13. Zhang Z, Meerow S, Newell J and Lindquist M 2019 Enhancing landscape connectivity through multifunctional green infrastructure corridor modeling and design
14. Baraya S (2019) Medellín creates 30 green corridors to mitigate urban heating
15. Briffett C, Sodhi N, Yuen B and Kong L 2004 Green corridors and the quality of urban life in Singapore https://cals.arizona.edu/pubs/adjunct/snr0704/snr07041g.pdf.

Index

J. T. Anderson (ed.), *Advances in Environmental Sustainability*, Springer Proceedings in Earth and Environmental Sciences,
https://doi.org/10.1007/978-3-031-26365-1